BEHIND THE GLASS

BEHIND THE GLASS

THE CHEMICAL AND SENSORIAL TERROIR OF WINE TASTING

Gus Zhu MW

ACADEMIE DU VIN LIBRARY

First published in 2024 by Académie du Vin Library Ltd
academieduvinlibrary.com

A CIP catalogue record for this book is available from the British Library
ISBN 978–1–913141–91–2

Publisher: Hermione Ireland
Editor: Rebecca Clare
Illustrations created by Ivy Xie
Cover image © Wirestock Creators/Shutterstock.com

Printed and bound in the U.S.A.

Contents

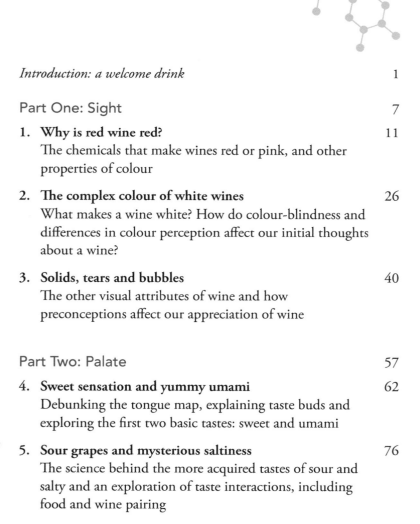

Introduction: a welcome drink

This is a book that explains the science behind wine tasting. Think of this chapter as the welcome drink, introducing to you what the book is about. As the title of the book suggests: behind every glass of wine, there is science. The science that we aim to explain in this book covers the chemical and sensory aspects of wine tasting.

The science behind tasting a glass of wine

I have often been asked to explain the science behind why a glass of wine tastes the way it does. There is no short answer, primarily due to two reasons. The first is that, chemically, wine is one of the most complex solutions in the world. While water and ethanol are its major components, a wine contains dozens of different acids, hundreds of higher alcohols, thousands of aroma compounds and millions of phenolic combinations. The second reason is that human senses are among the least understood subjects in the scientific world. Despite some substantial breakthroughs in fields like physiology, neuroscience and sensory science, we are still at the embryonic stage of understanding perceptions like touch, pain, temperature perception, hearing, vision, taste and smell. Moreover, human senses are complicated by the fact that we are all different from one another. The same glass of wine is perceived in different ways by different people.

In short, when you consider the potential chemical complexity of wines, alongside the fact that our own sensory systems are not well understood, it turns the seemingly simple question, 'Why does it taste that way?' into a surprisingly complex one. Despite this complexity, I am eager to show you how wonderful it is to indulge yourself in the chemical and sensorial world of wine. That is why I chose this complex area as the subject for my first book. To explain the science of wine tasting to readers I have developed two concepts: the chemical terroir and the sensorial terroir of wine.

Terroir: an evolving concept

The French term terroir has the literal meaning of 'soil' or 'a sense of place' but its true meaning goes far beyond this. Foods and drinks such as coffee, chocolate, tea, olive oil and tomatoes can vary in taste depending on where the crop was grown. However, in the wine world, such terroir expression is much more distinctive. For example, without having any background information, solely by assessing the liquid in the glass an experienced taster can pinpoint the identity of a wine, including its grape variety, region of origin and even its vintage. It is widely acknowledged that wines are much more closely associated with their terroir than other foods and beverages.

In the wine world, terroir is traditionally defined as a combination of environmental factors including soil, topography and climate that give distinctive characteristics to each glass of wine. Over hundreds, if not thousands, of years of wine history, the concept of terroir has been constantly evolving. For instance, a 'people' element is now often included in the concept of terroir, because human factors such as approaches to grape growing and winemaking have a great impact on how a wine expresses its sense of place. Recently, scientists advocated the concept of 'microbial terroir', as the microorganisms of a vineyard or winery can significantly affect the taste of a wine.

It is safe to say that the concept of terroir has become so broad that it can refer to any factors that shape the unique sensory profile of a wine. Hence, this book uses the concepts of chemical terroir and sensorial terroir to illustrate the elements that influence our tasting.

The scope of this book

Throughout this book we will divide our attention between two aspects of wine tasting: the chemical and the sensorial.

The chemical terroir of wine tasting is about the chemical composition of the wine itself that contributes to the sensory profile. Every glass of wine has a unique matrix of chemical components that forms a terroir of its own. Each chemical compound, as well as the combinations of those compounds, provides varied stimuli to our senses when tasting wines. With advances in chemical analysis, specialists such as flavour chemists have a better understanding of the chemical nature of flavour compounds and their corresponding sensory outcomes. Although the taste of wine is derived from natural fermentation and cannot legally be engineered, flavour chemistry and some other scientific disciplines benefit a lot from understanding the chemistry of wine. For example, winemakers can adjust the concentration of aroma compounds, such as certain sulphides, in order to encourage or discourage the grapefruit or passion fruit-like smell in wines made from the Sauvignon Blanc grape, according to the aroma profile demanded by the market. Therefore, roughly half of the mysteries behind wine tasting are explained in this book by the chemical terroir.

The sensorial terroir of wine tasting refers to human perceptions of the chemical compounds in wine. The focus is on the human senses rather than chemistry. There are two broad approaches to understanding human senses in the academic world. One is the instrumental analysis of the anatomy of sensory receptors, the nervous system and the genetics relevant to human senses. The other is sensory science, which uses humans to measure the sensory properties of products or to assess consumer preferences. All sensory-related fields are at an early stage of development due to the short history of these scientific disciplines and the rather recent advancement of the relevant measuring tools. Via state-of-the-art lab equipment, carefully designed experiments and sensible interpretation of statistics, we now understand more about how humans perceive wine and how varied people's responses to the same glass of wine can be.

As this is not a textbook (and there is no exam) there is no need to have foundational knowledge of chemistry and sensory-related fields in order to appreciate this book. I will summarize and explain the results of scientific research as we explore the subject so there is no need to understand the methodologies and data analysis behind the studies I refer to. It is my intention that this book will serve as a reference for everyone during wine tasting, regardless of their scientific knowledge level. This is not a book to teach you specific ways of wine tasting, either.* It is to help clarify the questions you might have during wine tasting, such as what it is in wine that causes a certain sensory perception when tasting it and why you may perceive this wine differently from another person.

The structure of this book

This book is structured according to the chemical and sensorial terroirs of wine tasting. Each chapter focuses on a specific category of sensory attributes as communicated in the wine industry or our daily lives. A chapter typically starts with a discussion of how certain chemical compounds contribute to those sensory properties, followed by a specific sensory theme which is closely related to the topic of that chapter. For example, when discussing the colour of wines, the chemical terroir is centred around the compounds that reflect specific wavelengths of light into our eyes, while the sensorial terroir covers topics like colour-blindness, which explains how the same wavelengths are perceived as

---

* Some established methods of wine tasting include the *WSET Systematic Approach to Tasting®* and the *Deductive Tasting Grid* by the Court of Master Sommeliers.

different colours due to our genetic differences. There are graphics throughout to help you understand certain elements of chemical and sensorial terroir.

The book is organized in three parts, in increased difficulty of sensory experience for most people. From childhood to adulthood, certain senses are better trained than others. Vision is our most well-trained sense, while our sense of smell is the least trained. For instance, most people can easily tell a red wine and a white wine apart visually (one is red, one is not!), but find it much harder to describe what a wine smells like. Therefore, the book begins by discussing how people appreciate wine visually, using their best-trained sense. It then explores the taste of wine on the palate, taste being generally less well-trained than sight. The final sense we explore is smell, as the book delves into the aromas of wine. This aspect of wine tasting can be intimidating to many inexperienced wine consumers because most people today do not rely on their sense of smell as much as they do their other senses.

In the last part of the book, I use six pairs of wines to summarize the concepts presented in all the previous chapters. Through those wine examples, you will be able to put the information and principles covered in this book together, whether you have access to those wines to taste or not. Hopefully, by the end of the book, the chemical and sensorial terroir of wine tasting will become crystal clear to you, and give you the knowledge to savour more layers of beauty in wine.

PART ONE
Sight

Drink to me only with thine eyes,
And I will pledge with mine;
Or leave a kiss within the cup,
And I'll not look for wine.

Ben Jonson, 'To Celia'

Observing with our eyes is a fundamental skill that humans are trained in from birth. For example, we are taught to match a red or square object with another object of the same colour or shape. Even blind people use touch and other senses to 'picture' the shape, size and weight of an object in their minds in order to better understand it. Much of civilization, development and entertainment is centred around our sense of sight. Therefore, the most intuitive part of wine tasting is the appearance of wine in a transparent glass. Sometimes, we believe in what we see so much that our vision can trick us. It is easy to ignore the fact that we all see things differently and the same object may show variations in its appearance in different conditions. In wine tasting, it's not uncommon to have arguments or confusion when discussing the appearance of a wine due to the discrepancies between what multiple observers see. The colour and the other visual effects of a wine give us a first impression. Before we become more intimate with the wine by assessing taste and smell, the visual clues play an important role in determining whether we like it or not. Therefore, this book, along with most books on wine tasting, begins with the appearance of wines. As well as the colours of wine we will note other visual characteristics such as the sediments, tears or bubbles that feature in certain wines.

Chapter 1 starts with the basic theory of colour, exploring our perception of colour through the wavelengths of light received by our eyes. This is followed by the introduction of a large group of chemical compounds called phenolics which produce multiple sensory attributes in a glass of wine. A detailed discussion on anthocyanins explores the key phenolic compounds responsible for the colour of red wines. Three physical properties of colour are presented to explain the pinkish colour of rosé wines as well as other complexities of colour. The last part of this chapter is about the colour receptors in our eyes, as we begin the journey of exploring the sensorial aspects of our vision.

In Chapter 2 we focus on the other types of phenolic compounds that contribute to the variations of colour in white wines. A preliminary discussion on the oxidation of phenolics explains the chemical reactions that contribute to the browning of white wines. At the end of the chapter, colour-blindness and the variabilities in normal colour vision are examined to showcase the extent to which genetics can lead to differences in our vision.

Chapter 3 explains the chemical and physical processes behind the visible characteristics of wine beyond colour. It demonstrates how solids, tears and bubbles are formed in a wine and how they are perceived through our eyes. The chapter concludes by giving examples of how psychological factors like our preconceptions affect the hedonic response when looking at a wine.

1
Why is red wine red?

The physics and chemistry of colour

Colour is the result of light being reflected and perceived by the human eye. Different wavelengths of light are perceived as different colours by our vision. When an object appears white, it reflects all the visible wavelengths of light into our eyes. On the other hand, when an object appears black, it absorbs all the visible wavelengths of light. All wines absorb light and reflect specific wavelengths of light depending on their chemical composition, resulting in their distinct colours. The simple explanation for the colour of red wines is due to the reflection of light in the wavelength range of red. This also means that red wines *absorb* other wavelengths such as those that appear as green colour.

Among the large population of chemical compounds in nature, certain compounds possess specific configurations of electrons that absorb and reflect photons, which are the elementary particles that we perceive as light. As a deeper look at chemistry and physics is beyond the scope of this book, what we really need to understand is: what are the chemical compounds in red wines that absorb the 'not-so-red' wavelengths and reflect the reddish wavelengths of light? The answer is that they are a group of compounds called anthocyanins, which belong to a large category called phenolics. In this chapter, we will see an entire chemical world centred on anthocyanins. Because of the many forms and chemical properties of anthocyanins, the colours of red wines are diverse and their colours change during the ageing process. This chapter becomes more colourful when the concept of colour properties is introduced. Even though the composition of anthocyanins in a glass of red or rosé wine is fixed, the alteration of any property of colour will lead to a difference in our perception.

Phenolics lesson 1: anthocyanins

Phenolics, or phenols, are ubiquitous in the plant kingdom and have thousands of different structures that correspond to myriad biological functions. Other than acting as building blocks for colour compounds in plants, phenolics play vital roles in forming structural integrity, assisting wound healing and repelling harmful insects and microorganisms. The human diet also includes plant phenolics, which are essential for better health, since we do not have the ability to synthesize phenolics in the body. Apart from colour, there are other sensory attributes based on phenolic compounds, such as the bitter or astringent taste on the palate. Without phenolics, wines would be chemically much simpler and, sensorially, much less interesting.

In nature, many living things have certain colours for a reason. To put it more precisely, organic life possesses compounds that give off certain wavelengths while absorbing others and those wavelengths are tools for attracting or hiding from others. For instance, male peacocks use their colourful tails to charm female peahens. It is also hypothesized that the black and white stripes of zebras provide camouflage from predators by creating visual illusion and confusion in grassland.

The key compounds that are responsible for the red to purple colour in grapes (and thus in wines) are anthocyanins. Recent genetic research indicates that all grapes were 'black' varieties in ancient times, meaning their skins (or even flesh in some varieties) were always red to purple in colour and rich in anthocyanins. The purpose seems to be obvious: such colour is a sign of ripeness in the eyes of animals who will eat the fruit and help the grapevine to distribute its seeds. But almost all plants have evolved other ways of reproduction, such as being propagated via cuttings.

Through evolution, some grape varieties may have emerged with 'silenced' genes responsible for synthesizing anthocyanins, yet still were able to reproduce perfectly well. Eventually, they became white-skinned, pink-skinned or brown-skinned varieties. In red and rosé winemaking, the anthocyanins in the skins of black grapes are released into the juice prior to or during fermentation. Essentially, winemakers 'taint' the juice and wine by incorporating the anthocyanins from the grape skins into the juice and resultant wine. So the key factors that contribute to a red wine's colour are the natural anthocyanin composition in the grapes and the level of extraction during winemaking.

The many faces of anthocyanins

'Anthocyanin' refers to a group of compounds sharing the same type of chemical structure. Rather than delving into the complexities of this structure, in order to appreciate their effect on a wine it is enough to understand that anthocyanins have the chemical property of absorbing green light and reflecting the red, blue and purple wavelengths of light, hence the red-to-purple colour perceived in our eyes. But how can we explain the variations in the colour of red wines? For instance, some wines are close to what we might call red, while some look a bit orange and others seem to be more purple in colour. In the wine industry, colour-related descriptions such as ruby, tawny and purple note the variations of colour in red wines. It is not about a pure, absolute colour called red. Chemically, one of the reasons for the variability lies in the different forms that anthocyanin molecules can take.

There are three forms of anthocyanin based on the chemical environment in wine: the flavylium form, the quinoidal base form and the carbinol pseudo-base form. Each form absorbs and reflects light differently:

- The flavylium form gives the red hue.
- The quinoidal base form absorbs and reflects slightly different wavelengths of light resulting in a blue to purple hue.

- The carbinol pseudo-base form lost its ability to absorb light and is therefore colourless.

Note that around 90 per cent of the anthocyanins in wine are in the carbinol pseudo-base form (colourless), but the rest are sufficient for contributing to the range of colours we call 'red'.

The three forms are pH-dependent, meaning the proportion of each form of anthocyanin present depends on the acidity or alkalinity of a substance. Flowers and fruits can alter their colour by changing the pH environment within them. For example, a Brazilian native plant, commonly known as the 'yesterday-today-and-tomorrow' flower, can appear as purple yesterday, turning less purple and more lavender-coloured today before becoming white after a few days. The mechanism of such rapid colour changes within the flowers is a change of pH that influences the colour of anthocyanins.

In wine, the general pH range can vary between 3.0 and 4.0. This determines the percentage of each form of anthocyanin present.* Lower pH (more acidic wines) results in more of the reddish flavylium form of anthocyanins, which partially explains why low pH wines, such as quite a few northern and central Italian red wines, often look ruby in colour. Higher pH (less acidic wines) increases the percentage of anthocyanins in the blue-purple quinoidal form, which suggests why higher pH wines such as those made from Syrah and Malbec tend to display a purple hue.

To understand the influence of pH on the forms of anthocyanin, there is also a little experiment we can do at home. Take a red wine or squeeze the red juice from red fruits such as strawberry, raspberry or cranberry, then dilute the wine or juice with water. The pH becomes higher because such dilution results in a decrease in acidity in the solution due to the higher pH of water (tap water has a pH of around 7.5). As the water is added,

---

* Different references present slightly different pH ranges, see more discussions on acidity and pH in Chapter 5.

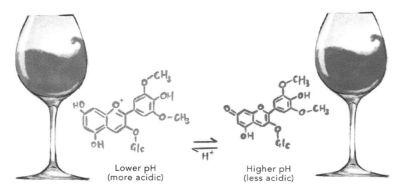

Figure 1: The colour of anthocyanins influenced by pH

the juice will turn from red to blue-purple in colour, reflecting the switch from the flavylium form to the quinoidal form of anthocyanins. If water is added further, we may see a significant loss of colour; not only because of dilution, but also due to the loss of the flavylium or quinoidal form of anthocyanins that contribute to the colour.

As we can see, anthocyanins need to be in a particular chemical environment in order to show colour. In other words, it is quite easy to lose the red colour if conditions are not right for anthocyanins. Besides being pH-dependent, anthocyanins can also react with certain chemical compounds and get bleached. During winemaking or bottling, if the producers add too much sulphur dioxide (SO_2), the colour of the wine can be lost. This is called bisulphite bleaching and comes from the interaction between the sulphur dioxide added and the anthocyanins. Bisulphite as the major form of sulphur dioxide in wine reacts with anthocyanins and turns them into colourless molecules. Paying attention to the sulphur dioxide levels in wine is crucial if a winemaker is concerned about the loss of colour in red and rosé wines.

White wines can also contain a low level of anthocyanins if they are made from black or pink/brown-skinned varieties. For example, the pink- to brown-skinned Pinot Gris/Grigio grapes can be made into white or rosé wines depending on the length of skin contact during winemaking. Even if a white wine has a yellow, golden colour at the beginning, some pink colour can be picked up as the wine is exposed to oxygen. The reason for this is that while the sulphur dioxide added during production bleaches the anthocyanins, oxygen can consume the sulphur dioxide to reverse the bisulphite bleaching effect, resulting in the return of anthocyanins to the coloured forms. Known as the pinking phenomenon, this is an unexpected and undesired change for makers of white wines.

Wine pigment: a union of anthocyanins

While there is no true consensus, chemists generally define wine pigment as condensed anthocyanin and tannin structures. Hence, at this point, we need to give a brief introduction to the other major group of phenolics in red wines – tannins. Grape skins and seeds have abundant tannins, which are extracted into wines to greater and lesser degrees during production. The tannins in grapes consist of a class of chemical compounds called the flavan-3-ols. In terms of chemical structure, tannins always consist of several to many flavan-3-ol units (for example, a longer, repeating chain of a given molecular structure), since these molecular units would be unstable if individual units were not linked together. Such chemical bonding is called condensation or polymerization (see detailed discussions on tannins in Chapter 7). Unlike anthocyanins, tannins by themselves are colourless.

What's more relevant to red wine's colour is that the chemical structures of flavan-3-ols look strikingly similar to those of anthocyanins, since they all belong to the same group of phenolics, called flavonoids. These structurally similar molecules are like siblings. Thus, a free anthocyanin molecule has the tendency to form a bond with an established tannin unit, creating wine pigment – a condensed anthocyanin-tannin structure that has a red colour. Therefore, when we see the red colour of a red wine, it usually consists of individual anthocyanin units as well as the anthocyanin-tannin combinations. When we taste the astringent tannins on our palate (to be discussed in Chapter 7), quite a few of the tannins will have been 'painted' with colour materials called anthocyanins.

As wine ages, the condensation of anthocyanin and tannin units continues, resulting in longer-chain, more polymerized wine pigments. Some wine pigments get large enough to become insoluble and eventually form sediments in red wines. What's happening here can be counterintuitive. As a red wine ages, the colour becomes lighter as some of the wine pigment precipitates in the bottle. Sensorially, we tend to perceive lighter coloured objects as dilute and weak. Chemically, however, it is quite the opposite. Those wine pigments, whether they are still dissolved in wine or have become solids, are much denser and stronger in the polymerized form.

The real world is always more complex than theory. Wine pigments are not just about condensed anthocyanin-tannin compounds. There is another class of compounds called pyranoanthocyanins which are formed by reactions between anthocyanins and other wine constituents. For our purposes, we need to know that these are very stable colour compounds. For example, a type of pyranoanthocyanin called portisin is found in Port. These portisins are among the wine pigments that give most Port wines a deep and stable colour. The existence of portisins, along with the condensation of anthocyanins and tannins, helps Port retain its deep colour over years or even decades.

Wines without prolonged ageing can also have colour stability, including the younger styles of Port. This is another phenomenon called copigmentation, which is a more transient interaction between anthocyanins and some other phenolic compounds like flavonols (a class of flavonoids). Such interactions enhance the colour and can even alter the wavelength absorbed by these 'crowds' of anthocyanins plus other phenolics. The result can be a more intense colour in red wines, as well as a tendency to be more purple in colour, just like many youthful red Port wines. Certain

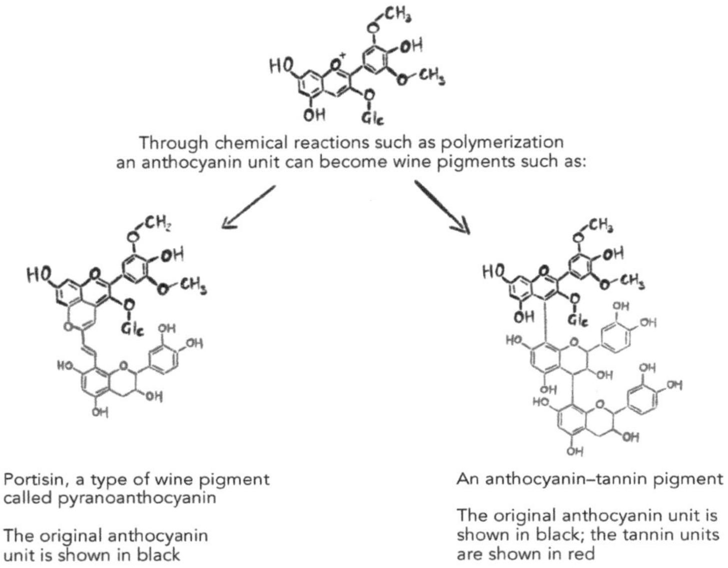

Through chemical reactions such as polymerization an anthocyanin unit can become wine pigments such as:

Portisin, a type of wine pigment called pyranoanthocyanin

The original anthocyanin unit is shown in black

An anthocyanin–tannin pigment

The original anthocyanin unit is shown in black; the tannin units are shown in red

Figure 2: Other forms of red wine pigment

grape varieties possess specific combinations of anthocyanins and flavonols that enhance the copigmentation. For example, Port varieties like Touriga Nacional tend to showcase intense purple colour due to the strong copigmentation effect. Anecdotally, it is believed that the Viognier grape variety adds copigmenting phenolics that enhance the colour of Syrah if the two varieties are fermented together. However, research has shown no evidence of improved colour stability, since Viognier (at least for the Viognier wines used in those studies) does not contribute to a higher concentration of flavonols that would improve copigmentation.

Some red wines, such as those based on the Nebbiolo grape, become more yellow, garnet or tawny in colour as they age. The exact reason is not well understood, but there has been evidence of newly formed wine pigments that reflect wavelengths that shift towards the orange–yellow zone in the colour spectrum. For example, the condensed anthocyanin-tannin compounds can be restructured further to become what's called the xanthylium pigment, which is more yellow in colour. In terms of the chemistry of wine ageing, research still has a long way to go.

This concludes our first lesson in phenolics. In Chapter 2, we will move on to Phenolics lesson 2: non-flavonoids.

Colour properties

So far, our discussion has focused on the chemical origins of colour. Hence, we simplified the sensory perceptions of colour to absolutes such as 'red' and 'purple'. But our life experience tells us that there are many different variations of the same class of colour. If we take two glasses of the same red wine and view them under different lighting conditions, one can appear to be brighter in colour, while the other is darker. Colour is never one-dimensional and multiple factors can tweak the appearance of colour. To explain the variations of colour in wine, three important colour properties are highlighted: hue, saturation and value. We'll consider each of these in turn.

Hue

Hue is used to describe the basic categories of colour such as red, green and blue. When we talk about changing hue, we are referring to a change between distinct colours; not a change in other parameters such as brightness or darkness. As explained previously, we can see a given colour when a specific wavelength of light is reflected into our eyes. But it is important to keep in mind that the wavelengths on the colour spectrum are continuous, meaning that within the range of 'red' coloured wavelengths, some wavelengths may appear closer to the colour we call red, while others will look more orange. When we looked at the chemical composition of colour we mentioned the shift of reflected wavelengths as anthocyanin-tannin restructured to become xanthylium in certain aged red wines, this modified the hue from 'more red' to 'more yellow' as a consequence.

Here, some numbers may help provide more intuitive understanding of the minor change of hues. The range of wavelengths in the visible light spectrum is from 380 to 700 nanometres (nm). As the number slowly increases or decreases, the transition of the corresponding colours is gradual. For example, a specific wavelength between 625 to 750 nm gives the red colour, while 590 to 625 nm represents the zone of orange colour. This means that the wavelength of 625 nm, which sits at the meeting point of the two ranges, can appear as a red-orange colour. Later in this chapter, we will discover the mechanisms that determine how our eyes perceive these wavelengths of light.

The reality is that there are often multiple different wavelengths of light being reflected by an object. As we have seen in a glass of red wine, there are multiple forms of anthocyanins and anthocyanin-derived compounds. Hence, the resulting colour is a mixture of wavelengths, whether the range is narrow or broad. The best example is a red wine that appears both ruby and purple in colour. Such a blend of colour is likely due to the purple-coloured anthocyanins in the high pH environment of the wine, along with anthocyanin-derived pigments that reflect wavelengths in the more reddish colour spectrum. The same theory explains wine with a ruby to garnet colour: these wines contain colour compounds that reflect wavelengths with higher and lower numbers (they can co-exist in the same pH environment) in the red-to-orange spectrum.

Saturation

Although the hue helps anchor the basic categories of colour, this property alone cannot explain other observations of colour. By looking at the visible light spectrum, you may find a popular colour that is missing: pink. Rosé wines, favoured by markets worldwide nowadays, attract consumers through their various shades of pink. Winemakers create such lovely colours by releasing a low level of anthocyanins from grape skins into the wine. It seems obvious that low levels of anthocyanins account for the pink colour in wine. But still, the colour property of hue alone does not explain the colour pink, which is non-existent in the visible light spectrum.

In order to understand this, another property of colour needs to be introduced: colour saturation. Saturation is how pure or brilliant a colour is. The alternative term for saturation is intensity, which is often gauged when observing a glass of wine. When saturation is reduced, the colour is more diluted, with a white tone. Although there is no wavelength of light with the hue of pink, when white colour is mixed with red, our brains interpret this as the colour pink (more discussion on this later). When a wine has a low concentration of anthocyanins, the red colour becomes weak. Therefore, other perceived wavelengths from the environment can have a significant impact on what we see. In this case, the light with all or most wavelengths reflected from a white background goes through the wine glass and cannot be overpowered by the small amount of red wavelengths in the wine. The result is that our eyes perceive a mixture of red wavelengths and the white ambient light to produce what we see as pink. In technical terms, rosé wines are pink because of the white background added to them, causing the saturation of the red hue to be decreased.

One of the implications of the pink story is the importance of the surrounding light while observing an object. In most cases, the

Figure 3: Colour property – saturation. As the saturation of colour decreases, the colour becomes less red, more pink, and whiter.

ambient light is white in colour, so that the saturation of colour is always lower when our eyes receive strong natural light. Thus, looking at a rosé wine against a shiny white background makes it appear pinker in colour. On the other hand, a rosé wine seems to be darker and more reddish in colour if you observe it against a piece of black paper. Given that colour is an important factor for the way in which consumers make decisions about a purchase, it is crucial to consider the lighting conditions and the ambient colours when displaying a product, for example if using fluorescent lights as the preferred source of lighting in a shop.

Value

Value is the brightness or darkness of a colour. As the value gets lower, the colour becomes darker and duller (greyer or with more shadow). The value of colour is influenced by the degree of light reflected from an object. The more light an object absorbs, the lower the colour value. On the other hand, if light goes straight into our eye without losing any power, the colour value will be at its highest. Saturation, discussed above, is about how much other light is mixed in to dilute the purity of the hue, whereas value is about the intensity of each hue.

When we check on the colour of red wine in a glass, such as by tilting the glass at a 45-degree angle and looking through the liquid in the glass against a white background, the common observation is that the centre of the wine appears to be darker than the rim. This is because the centre has more wine and therefore a higher concentration of compounds (more than just anthocyanins) that 'devour' the light, resulting in lower value and more darkness of the colour when compared to the rim, with less concentrated light-absorbing compounds. If you care about the observation of colour in wine tasting, it is important to have sufficient natural light. Tasting in an old, underground wine cellar can be atmospheric, but the dim lighting in those places makes the colour of wine irrelevant, as everything looks grey and dark. Once again, the ambient light conditions have a significant influence on the properties of colour.

Thanks to some magical software and electronic devices, these days we can understand colour properties more easily and intuitively without necessarily understanding the physics involved. Graphic editors on our computers and phones give us access to colour manipulation, enabling us to adjust various parameters of colour. A glass of dark red coloured wine in a photo turns pink as you lower the level of saturation. As the value of colour is edited to a high level, the colour looks very bright. The result is a shiny, rosé wine that conceals the 'true' colour. When shopping online, be careful when you look at the colour of a dress based on the images shown on the screen. If the colours in the photos have been heavily modified (or even due to the difficulty in calibrating a computer screen to match reality) you may receive a dress of a significantly different colour.

Colour receptors in the eyes

So far, our discussion has been focused on the physical properties of light and colour, without really examining the physiology of human vision. Since this book also explains the sensorial terroir, we need to look at how exactly our eyes perceive light and how our brain interprets those wavelengths.

Human eyes have colour receptor cells called cones. There are three types of cones in the retinas of our eyes: the L-cone (L for long), the M-cone (M for medium) and the S-cone (S for short). As the name suggests, the L-cone is more sensitive to long wavelengths, in the red–orange colour zone; the M-cone is more sensitive to medium wavelengths, in the yellow–green colour zone; the S-cone is more sensitive to short wavelengths, in the blue–purple colour zone. A given wavelength hitting our eye might fully trigger one type of cone, partially trigger another or not trigger another cone at all. Here is a list of examples to help you get a feel for how the cones work:

• When a long wavelength of 700 nm hits our eyes, the L-cone gets fully switched on, whereas the other two cones don't respond to it at all. The result is a pure red hue being perceived.

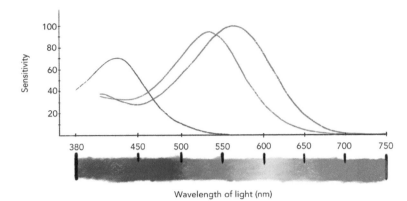

Figure 4: Colour receptors for different wavelengths

- If the wavelength is switched to around 570 nm, both the L-cone and the M-cone are highly activated, but the S-cone is still inactive. At this moment, the lively L- and M-cones are translated into the colour yellow by our brains.
- Now say the wavelength becomes shorter, to around 500 nm. The L-cone becomes less active and the M-cone dominates, while the S-cone is still sleeping. Our brain interprets this as the colour green.
- If the wavelength is a short one, of 450 nm, the red and green cones are less sensitive but still partially responsive, while the S-cone is finally awake, resulting in the perception of the colour blue, or even slightly purple.

Common sense tells us that our vision should perceive more than just the different types of hues. As indicated by the colour properties of saturation and value, colour intensity and background lighting also affect the colours we see. Indeed, our vision does contain another group of receptor cells responsible for sensitivity to light. Those receptors are called rods, and they are crucial in our daily lives, for example, in giving proper vision at night when the light is weak. Rods also have a prime zone of wavelengths, which is around 500 nm, corresponding to the green to blue colours. No wonder most objects look dark green and dark blue in the evening.

Final thoughts

From anthocyanins to colour properties, I hope the first chapter of this book opened your eyes to the kaleidoscopic world of chemistry regarding the colour of red and rosé wines. In the short section on colour receptors in our eyes we only took a peek into the sensory world of human vision. In the next chapter, we will broaden the view by examining the colours of white wines and by understanding how differently two individuals may perceive the colour of the same glass of wine.

2
The complex colour of white wines

Is white wine 'white'?

If a glass of white wine and a glass of milk are viewed side by side, milk will obviously win the 'which is whitest' competition. The more appropriate interpretation of the colour of most white wines should be 'almost transparent and water-like'. A white wine does not have that many chemical components that reflect all the light in the visible spectrum to be perceived as white as milk. Instead, it contains low levels of colour compounds so that most of the lighting penetrates through the clear wine glass and the wine. Because of this, the background colour, as we discussed in Chapter 1, has much more influence on the appearance of a white wine than it does for red wine. This is also why a truly white background like a piece of white paper is better for assessing the colour of wine.

No wine is crystal clear in the way that pure water or vodka are, because most white wines have many different types of chemical compounds that contribute to perceivable colours, even if those compounds are in small concentrations. The broad range of colours of white wines is reflected in the many descriptors that are often heard in the wine trade. If a white has some green hues, it can be described as straw, lemon or lemon-green in colour; if the hue is distinctly yellow, the common descriptors include gold, golden or simply yellow; if a white wine has experienced some degrees of oxidation, it might be called amber, copper or brown in colour.

In this chapter, we will start with the compounds of chlorophylls and carotenoids which contribute to the green and yellow hues of white wines. The second lesson on phenolics will help us to understand the browning phenomenon observed in many plant-based products, including white wines. The last part of this chapter delves deeper into human vision. In particular, colour-blindness and variations in normal colour vision are explored to remind us how varied our perceptions of the world can be.

Chlorophylls and carotenoids

You might be familiar with these well-known types of colour compounds, since chlorophylls are responsible for the green colour of leaves and carotenoids produce the orange colour in carrots. What you may not have realized is that we can also find these compounds in wine. The green hues in white wines often come from low levels of chlorophyll in the grape skins or from vine leaves. Certain varieties, such as Sauvignon Blanc, have some chlorophyll in the skins even when the grapes reach full ripeness. In addition, the harvesting of grapes often includes materials other than grapes, such as vine leaves, which may be included (both deliberately and accidentally) upon crushing and pressing during production. Those leaves also release a small amount of chlorophyll into the grape juice to be fermented.

Chlorophylls are crucial compounds for photosynthesis. They are responsible for absorbing light, from which all living things on planet Earth derive their energy. As we learned in Chapter 1, our eyes receive the wavelengths of light that are reflected by an object. Thus, it is not hard to understand that chlorophyll absorbs red and blue light, but not green light, hence the green colour of vine leaves. Scientists have proposed theories about why chlorophylls select specific wavelengths, but definitive proof of the exact reason remains elusive. One leading explanation is that many genes of modern plants were taken from green algae that lived under the ocean hundreds of millions of years ago. Those microorganisms only had access to red and blue light penetrating through the deep water. Consequently, chlorophyll developed the ability to capture the available wavelengths in those two ranges, but not in the zone of green wavelengths. Even today, almost all consumable energy for organic life on Earth comes from the process of photosynthesis in plants, algae and certain bacteria. Without chlorophyll, no organic life would exist.

Carotenoids are a group of compounds that give a yellow hue to white wines. A vivid red, orange or yellow coloured carrot contains abundant carotenoids. Grape skins develop such compounds in lower concentrations, hence the pale lemon or gold colour rather than intense yellow of most white wines. Carotenoids belong to the chemical class of terpenoids which are ubiquitous in the plant kingdom and serve many roles.* Unlike chlorophyll, carotenoids absorb wavelengths in the range of green to violet light, leaving red to yellow light to be reflected. As a 'newer' compound than chlorophyll, carotenoids were created after some plants started to live above water with access to all the wavelengths of visible and non-visible light. Note that violet, blue and green light with their shorter wavelengths have more energy. Therefore, carotenoids choose to capture energy efficiently by absorbing the shorter wavelengths of light to conduct photosynthesis along with chlorophylls. Moreover, carotenoids do not blindly grab all light with short wavelengths. Too much ultraviolet (UV) light is damaging to most living things due to its higher energy. In order to protect the other parts of a plant, the carotenoids cleverly reflect the (invisible) UV light instead of seizing it.

Phenolics lesson 2: non-flavonoids

The colour in red wine (discussed in Chapter 1) and the astringent taste (to be discussed in Chapter 7) come from the direct sensory impact given by phenolic compounds like anthocyanins and tannins. In contrast, many white wines lack prominent colour

---

* In Chapter 8, we will explore how nature creates thousands of terpenoids with distinct aromas.

and an extra dimension of mouthfeel, so consumers are usually unaware of the presence of phenolics in those wines. However, without phenolics, some colours, along with many other sensory attributes of white wines would not exist.

Most of the phenolic compounds in white wines are colourless at the very beginning of the winemaking process. But some of them can go through a darkening or browning process. Such changes of colour in juice or white wines are observed frequently in fruits and vegetables in our daily lives. When we cut apples or white mushrooms, the phenolic compounds in the cross-sections are exposed to oxygen in the air. This begins the oxidation reaction and cross-sections turn brown in colour in a short period of time.

Chemically, the major phenolics for white wine's colour are the hydroxycinnamates, which belong to one of the phenolic categories called the non-flavonoids (whereas the anthocyanins covered in Chapter 1 fall into the category of flavonoids). The non-flavonoids are mainly found in grape juice and grape skins, but some can also come from wood if oak is used during wine production. This is clearly seen if we look at distilled beverages such as whisky. Whiskys start out colourless, with few phenolics, but they pick up a gold colour as they age in oak barrels and gradually turn a deeper brown as time goes by. This is because some phenolics in the oak wood barrel are extracted into the whisky and go through browning during ageing. To winemakers and consumers, the browning of white wine is a hot topic. Most white wines that are enjoyed nowadays are lighter-coloured examples. With the exception of a few special styles, a young white wine that becomes brown quickly is considered faulty. In wine production as well as in the processing of fruit and vegetable products, the prevention of browning is a major consideration. To understand the browning of non-flavonoid compounds in grape juice and in wine, we need to explore two mechanisms: chemical browning and enzymatic browning.

Chemical browning

A more familiar term for what happens during chemical browning is oxidation. As mentioned, the non-flavonoids like hydroxycinnamates in white wines are generally colourless. But once oxidation happens, those phenolics will be transformed into a group of compounds called quinones. Similar to certain flavonoid compounds discussed in the previous chapter, the quinone compounds will condense or polymerize, resulting in large pigments called melanins. Melanin is the exact substance that is responsible for the dark pigmentation in our hair and skin. In other words, the final formation of melanin through oxidizing hydroxycinnamates contributes to the browning in wine.

Three key elements are indispensable for the chemical browning process: the phenolics, oxygen and also metal ions, which have gained much more attention from chemists examining oxidation mechanisms in wine in recent decades. Even with abundant phenolic compounds and oxygen, without any metal ion present in the solution, a glass of white wine may not go through browning or any degree of oxidation over a long period of time. The metal ions, especially the two oxidation states of iron (the ferrous cation Fe^{2+} and the ferric cation Fe^{3+}) play a crucial role in catalysing the conversion from phenolics (non-flavonoids) to quinones and eventually to melanins.

Since browning is often treated as problematic during production and storage, winemakers will use compounds that are classified as reducing agents to inhibit or resist oxidation. The most common example is sulphur dioxide (in the form of bisulphite in wine) which bonds quickly with quinones, changing them back to the less oxidative state and preventing them from forming the melanin pigment. Therefore, the addition of the appropriate amount of sulphur dioxide during winemaking is important in terms of the prevention of browning during a white wine's short-term shelf-life.

It is worth emphasizing that when examining the issues of chemical browning, each of the three causes should be considered. For example, there is a strange phenomenon called premature oxidation (often called premox in the wine trade). Despite sufficient sulphur dioxide additions, many bottles of white Burgundy wines from the mid to late 1990s suffered from browning issues over a short period of time. One hypothesis is based on the fact that no matter how much sulphur dioxide is added, a high concentration of metal ions speeds up oxidation reactions at an exponential rate. With the presence of enough metal ions in wine, the formation of melanins can happen so quickly that it occurs before sulphur dioxide can bind with intermediate products of quinones to stop further oxidations. Coincidentally, in years with high fungal disease pressure back in the 1990s, some growers in Burgundy sprayed quite a lot of fungicides containing iron or copper ions onto the leaves and grapes. Therefore, it is possible that a higher level of metal ions in the grapes and wines was one of the culprits in the premature oxidation of Burgundian white wines from those years.

A few types of wines are intentionally crafted to be dark and brown in colour. These wines go through deliberate oxidation during winemaking to facilitate browning as the signature of their style. The increasingly popular amber wines, also known as orange wines, are made by extracting lots of phenolics, including tannins, through the fermentation of white grapes on their skins (just like red winemaking). Thus, these wines have a much larger pool of phenolics to be chemically oxidized to begin with. Even given a short period of maturation and ageing, the wines can easily become amber or brown in colour (or more orange in colour if the grape skins have some red-coloured anthocyanins, as is the case with varieties like Pinot Gris/Grigio). Long-aged wines with deliberate oxidation treatment like the oxidative style Sherries (e.g. Oloroso, PX, VOS and VORS) are distinctly brown or even black in colour. Some are matured for an average of 30 years to develop highly concentrated melanins that deepen the colour. It can be hard to believe that these wines started out as pale and gold-coloured grape juice. What are more fascinating about these deliberately oxidized wines are the wonderful aromas from the long oxidation process, which will be covered more in Chapter 12.

Enzymatic browning

In contrast to chemical browning, enzymatic browning involves organic enzymes as the catalyst for the oxidation process instead of inorganic metal ions. Enzymes are proteins produced by a living organism that accelerate a specific biochemical reaction. Plants and animals develop enzymes responsible for expediting browning as pigments like melanin for the purposes of self-defence. For example, the black-coloured melanins developed in our skins absorb not only all the visible light, but also the invisible wavelengths, such as UV radiation, which is harmful, hence the protection of our skin cells. Plants do the same to protect themselves. If we look at the freshly cut apple again, the enzymes in the fruit catalyse the browning reactions (i.e. they facilitate or speed up the reaction) so efficiently that the cross-sections exposed to the air develop dark pigments to protect the rest of the apple's flesh in a matter of minutes.

In juice and wine, the main types of enzymes responsible for browning are polyphenol oxidases (PPOs) and laccases and each type contains several species of enzymes. PPOs evolved to facilitate the oxidation of phenolics like hydroxycinnamates. These phenolic compounds will quickly be converted to quinones under the influence of active PPOs and subsequently condense to form melanins.

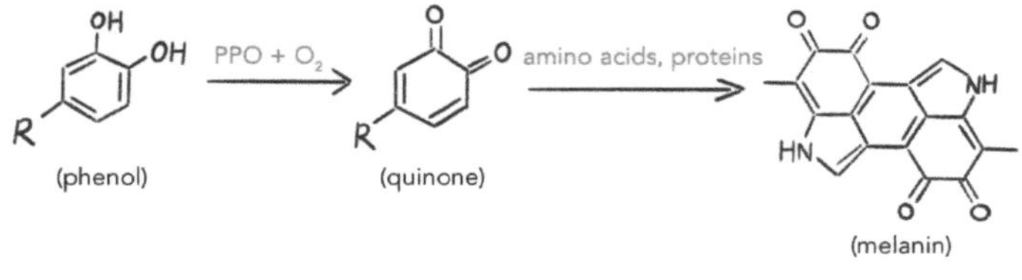

Figure 5: The enzymatic browning process. Enzymes such as polyphenol oxidases (PPOs) and laccases are catalysts for the enzymatic browning process.

Based on various research on grape juice, wine and other fruits or vegetables, we have learned that the optimum pH range for PPOs activity falls in the range of 4.0 to 6.0. This means that a more acidic wine with a low pH value around 3.0 can naturally inhibit the activity of PPOs. The other factor that can inhibit the PPOs is the addition of sulphur dioxide, which also reduces the formation of quinones as discussed previously.

The laccases, on the other hand, are much more troublesome in that they are very active even at lower pH levels. Additionally, laccases can tolerate relatively high levels of sulphur dioxide as well as being able to turn a much broader range of phenolic compounds into quinones and melanin pigments. Such troublesome laccases are often induced by the infection of moulds or other fungi such as the famous *Botrytis cinerea*.* When observing many bottles of classic sweet wines like Sauternes, Tokaji Aszú and Trockenbeerenauslese, it is noted that they have a tendency to turn brown easily despite the generally high dosage of sulphur dioxide during production. This is because these nectars are the famous styles made from *Botrytis*-infected grapes, which contain a significant number of laccases. Remember the premox issue in Burgundian white wines we mentioned earlier? Another hypothesis is that those wines which suffered from early-stage browning might have been produced from grapes that were slightly affected by pathogenic fungi, leading to an increase in the level of laccases in the grape juice.

It is worth mentioning that metal ions still play an important but indirect role in enzymatic browning. In the chemical structure of PPOs and laccases, the copper ion is an essential element that has a large impact on the functionality of these enzymes. For this reason, some health supplements include metals like iron and copper. The tiny quantity of metal ions activates and keeps

---

* This is informally called 'noble rot' by the wine trade since although these moulds attack the grapes, causing rot, in their benevolent form this can result in premium sweet wines, appreciated by nobility in the past.

the relevant enzymes in our body active, leading to a healthy metabolism. In the future, perhaps some commercially viable technologies will become available to control browning or the degree of either chemical or enzymatic oxidation in wine by manipulating the metal ions during wine production.

Colour-blindness

Here, our discussion on the chemical terroir of colour comes to an end. But sensorially, more topics remain to be explored and a major subject is the variation in human colour vision. Colour blindness is an extreme case of how much visual perception can vary between people. It is not uncommon to encounter someone who was born with colour-blindness. There are about 300 million people in the world with a deficiency in colour vision. It is estimated that 8 per cent of the male population and 0.5 per cent of the female population have varying degrees of colour-blindness.

Why are more men colour-blind? Genetic studies have shown that the genes that regulate red–green colour vision are heavily coded on the X chromosome. If there is an error in the genes on the X chromosome, that error can be passed down to the next generation. As women have two X chromosomes and men have only one, the chance of having two X chromosomes (in females) with the same error is lower than it is for men, who rely on only one copy of this chromosome. In other words, for women, if the X chromosome from one parent contains the genes of red–green colour-blindness (called 'deuteranomaly', see below), the X chromosome from another parent with genes of normal colour vision can help correct the problem. For men, the other half of the pair is the Y chromosome, which does not carry the genes that control red–green colour vision. In this case, unfortunately, the Y chromosome cannot be the potential backup to lower the chance of being colour-blind. Male readers of this book should not feel downhearted by understanding these facts, which merely demonstrate how genetics predetermine some of our senses, such as colour vision, and make us different from one another. There are four types of colour-blindness, which we'll cover in turn.

Deuteranomaly

Deuteranomaly is the most common type, and involves a reduced sensitivity to green light. Remember the cones, the colour receptors in our eyes, discussed in Chapter 1? In the case of deuteranomaly, the M-cone responsible for detecting medium wavelengths of the green light zone is less functional. Therefore, people with this type of colour-blindness perceive green as more orange in colour. However, people with deuteranomaly are not usually hindered by such vision defects in daily life. For example, the distinction between the green and the red traffic light is still obvious to those with deuteranomaly, even if the green light does not look so green to them. In wine tasting, it is only white wines with green hues that can be tricky for people with this condition.

Protanomaly

Protanomaly is the second most common type of colour-blindness and describes a reduced sensitivity to red light. In this case, the L-cone for long wavelengths of red to orange light is weak. People with protanomaly perceive red objects as less red and more blue, purple or grey (and they can still see a difference between the red and the green traffic lights). People with protanomaly find it troublesome when observing the colour of red or rosé wines because the colour of those wines appears greyish-green or blue and less vibrant to them.

Both deuteranomaly and protanomaly belong to the group of red–green colour-blindness. It is also worth noting that the two terms ending with '-anomaly' refer to a reduced sensitivity as expressed in the previous two paragraphs. But when the term deuteranopia or protanopia is used, it means almost complete blindness towards green light or red light. This means that the M-cone and/or L-cone of these people is almost completely disabled. People with deuteranopia cannot perceive the vivid green colours in nature, whereas those with protanopia do not recognize the colour in all red wines. (Neither group should drive as they are not able to distinguish traffic light colours.) Deuteranopia and protanopia are much less common than deuteranomaly and protanomaly among the human population.

Tritanomaly

Tritanomaly is a rare type of colour-blindness with a reduced sensitivity to the short wavelengths of blue and purple light. Less than 0.01 per cent of the population have this defect in their vision. In tritanomaly, the S-cone in the eyes has reduced sensitivity. People with this abnormality of vision often confuse blue with green and yellow with red or pink. Those with tritanopia (complete dysfunction of the S-cone) see a very different world than others due to the completely dysfunctional blue–purple light receptors. Wines such as those made from the Malbec variety tend to be distinctly purple in colour, but in the eyes of those with tritanomaly, Malbec wines look more blue-green. Note that men and women are equally affected by tritanomaly and tritanopia because the blue–purple receptor genes are on another chromosome (identified as chromosome 7) instead of on the X chromosome.

Monochromacy

The rarest and the most unfortunate case of colour-blindness is monochromacy. It involves the complete inability to distinguish colours due to the dysfunction of all three types of cones. It is estimated that only 1 in 33,000 people worldwide have this vision defect. The world is black, white and grey to those people, but they can still enjoy wine through the senses of smell and taste.

Lastly, Chapter 1 also addressed the rods that are responsible for sensitivity to light intensities. The small number of people with monochromacy (impaired cones) usually have functional rods, but anyone whose rods have become problematic will also have issues with their cones. Those individuals with impaired cones and rods have near-total blindness.

deuteranopia

protanopia

tritanopia

normal vision

Figure 6: Normal vision versus colour-blind visions

Variations in normal colour vision

In recent decades, an increasing body of research has shown that even among people with normal colour vision, colour can be perceived quite differently. For instance, light receptors like cones are made up of specific arrangements of amino acids determined by the genes. The most well-studied variation in vision is related to the two types of amino acids expressed by the variant of a gene in the L-cone. Located at a particular position –180 – in the chromosome, the difference in this gene among people can translate to the amino acids of either serine or alanine in the L-cone cells. Since the L-cone is responsible for receiving the longer wavelengths in the red-to-orange zone, different compositions of amino acids at position 180 between different people mean different sensitivities to red to orange lights.

In an experiment, a group of people were asked to adjust a mixture of red and green lights to match a standard yellow light. Those with the amino acid serine at position 180 used significantly stronger red light to complete the match than the ones with the amino acid alanine at the same position, demonstrating their different visual perceptions toward red light. It is estimated that among Caucasian males, about 62 per cent have serine at position 180 in the L-cone, while 38 per cent of them express alanine in that position.

In other cases, genetic differences in the light-sensitive rods can lead to variations in perception within specific lighting conditions. To those who suffer from poor night vision, some evidence has shown that supplementing more vitamin A can help (although further scientific proof is required to verify this). This is because vitamin A is an important element in the functionality of rods, as research over the last two decades has proved.

It is not just about the inherent genetic variations among people. Each person may experience different colour vision in various situations. There is evidence that physical fatigue and stress can cause temporary reductions in sensitivity to colour and light intensity. For example, several experiments have shown that when people were moved to much higher altitudes via climbing or transportation, more than 70 per cent of them proved to have a temporarily reduced sensitivity towards blue-purple light. This is likely due to the influence of the physical fatigue experienced at high altitude on the functionality of cone cells in our eyes. These observations imply that health status may enhance or deteriorate our colour vision. Many studies have suggested that in general, our ability to differentiate colour gradually deteriorates as we age, especially after the age of 70. It is also proven that patients with Alzheimer's disease tend to have a deficit in colour vision, and type 2 diabetes increases the risk of acquiring a colour vision defect.

Final thoughts

Before I entered the professional wine field, I assumed white wines would be simpler than red wines in terms of chemical composition. But the reality is that white wines are chemically very complex. The colours of white wines alone involve multiple compounds, from chlorophyll and carotenoids to various types of non-flavonoids. The chemical and enzymatic browning of white wines is even more intricate.

This chapter has started to explain how differently we perceive the world, by discussing colour-blindness and variations in normal colour vision. You will learn more about our individual variabilities when we explain the senses of taste and smell. For now, just keep in mind that the colour you see is what you see at a particular moment; as you get older, or if you simply become fatigued, your perception could change. This means that everyone's perception of colour (and thus their visual impression of a wine) is, to some degree, subjective.

3
Solids, tears and bubbles

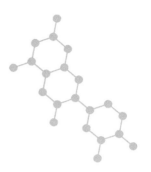

Beyond colour

As highly visual creatures, humans are not only drawn to certain colours but are also influenced by the other visual effects of our surroundings. For example, even the style of the glass can change our visual perception of a wine. A glass of wine is more than just a volume of liquid sitting a receptacle. It can include solids that make it appear cloudy, or dissolved gas, which gives the bubbly appearance upon opening a bottle. The extra layers of sensory attributes perceived by our eyes add to the diversity of wine styles and lots of fun during consumption. This chapter aims to explain these other aspects of a wine's appearance and their significance in our appreciation of a wine. We will refer to both chemistry and physics in understanding these unusual aspects of wine tasting, but do not worry – you don't need to do complex equations or run calculations in order to understand the relevant mechanisms at play.

Beyond colour, there are three other broad categories to observe in wine: solids, tears and bubbles. I will illustrate the chemical terroir of each category in detail. The chapter ends with a discussion of how our preconceptions affect our sensory perception and wine appreciation.

Solids

During the processing of most beverages, many different types of solids form, shaped by different chemical compounds. The solids formed in the production of certain drinks may be unpleasant or even unsafe to consume, which is why the majority of beverages go through a clarification process such as filtration. A newly made wine always contains some solids, which are generally safe to consume. In most cases, though, those solids in wine will be removed by fining and filtration upon bottling due to consumer expectations. We'll talk more about the influence of expectations and preconceptions later; first, let's explore the major types of solids and their corresponding chemical causes.

Polymerization of phenolics

In Chapter 1, we explored the polymerization of phenolics as the reason behind the accumulation of sediments in aged red wines. It all started with the two broad categories of phenolic compounds in wine that contribute to significant sensory qualities: anthocyanins (responsible for the colour of red and rosé wines) and tannins (contributing to the astringent mouthfeel). The chemical reaction called polymerization was also explained in Chapter 1: anthocyanins and tannins have a tendency to combine or condense to become more stable wine pigments. More polymerization of the phenolics means the molecule sizes of wine pigments become larger. Eventually, the wine pigments are large enough and settle down as red-coloured, visible particles (sediments).

Other types of phenolic compounds such as quercetin may also polymerize and precipitate as sediments, especially in wines made from grape varieties such as Sangiovese. Due to the lack of anthocyanins, the polymerized quercetin sediments are often colourless but can appear as white flakes depending on the lighting. In the eyes of most consumers, white flakes in red wine look unpalatable and undesirable. Producers of Brunello di Montalcino in Tuscany, Italy (Brunello is the local name for the Sangiovese grape variety) have identified quercetin precipitation in their wines and have been looking for the best practices to eliminate this sort of problem.

Lees

Winemaking includes the crushing and pressing of grapes along with a small amount of other solids (such as dust and pomaces – the pulpy residues remaining after grapes have been crushed) from the vineyard. Solids are also formed during fermentation; these include the mass population of dead yeast cells as well as insoluble components from the grape must (freshly crushed grape juice that contains the skins, seeds and/or stems). After fermentation, these solids agglomerate together to become sediments, called the gross lees. Winemakers may rack off (a way to remove) the gross lees to gain much clearer wine for further maturation or storage prior to bottling. After this process, there are still some solids left behind. Known as fine lees, these consist largely of dead yeast cells. Without the processes of filtration and fining, wines will look cloudy due to different levels of lees.

Certain wine categories, especially those qualified as natural wines, can contain lees that appear as sediments or cloudiness. These wines deliberately receive zero or minimal fining and filtration during winemaking. But even clear wines can turn cloudy due to microbial activities, which mainly include refermentation by yeasts or malolactic conversion (a process that converts the malic acid in wine into lactic acid) by bacteria. For example, in the past, white wines from the Vinho Verde region of Portugal were bottled with residual sugar. If there were some viable yeast populations remaining in the wines upon bottling, the result was a greater chance of refermentation by yeasts in the bottle. Such uncontrolled refermentation often causes a glass of wine to become hazy, fizzy and even in some cases unsafe to consume.* Any viable lactic acid bacteria in the bottle can consume malic acid (if any is present) to amplify their population and also cause turbidity. Avoiding such problems in today's winemaking world is crucial, except for those wines that are designed to have some chemically stable lees as sediments in them.

---

* These days the winemaking process for Vinho Verde is much more sophisti-
  cated and therefore the wines are not at all unsafe.

Tartrate crystals

Bottled wines, especially white wines, can form visible, crystal-like sediments. The compound responsible for such a crystallizing phenomenon is potassium bitartrate. In wine, potassium bitartrate is often supersaturated, meaning that the wine is at its full capacity in terms of its ability to dissolve the molecule. If any further potassium bitartrate is added, or if conditions change to further reduce the wine's ability to dissolve molecules (for example, if the wine gets colder), potassium bitartrate will become supersaturated. Since the wine solution cannot dissolve the entire population of potassium bitartrate any longer, some of it must precipitate. When the temperature becomes very low, it is therefore possible for a proportion of the potassium bitartrate to become ice- or crystal-like solids, and the wine has a reduced 'burden' of dissolving all the potassium bitartrate.

The crystallized potassium bitartrate is harmless because it is part of the wine. However, consumers may perceive these crystals in the bottle as broken glass. Thus, winemakers use various methods such as cold treatment upon bottling to minimize the risk of crystal formation in bottled wines. The wine is treated to force more potassium bitartrate to precipitate and the resulting crystals are removed by filtration. The wine is then less likely to become supersaturated with potassium bitartrate.

Protein haze

All wines contain a low level of proteins. These proteins are generally dissolved in wine so that our eyes cannot perceive them but under certain conditions they may become visible. Think about when we cook eggs: the raw egg white looks like a clear liquid, but as the temperature rises, the chemical nature of the proteins in the egg white is changed and it becomes opaque white in colour. In wine, the mechanism is the same. If a bottle of wine is stored at a high temperature, the proteins in it become insoluble. The sensory outcome is that white, glue-like matter will appear in the bottle, which looks very unappealing. This can be seen more clearly in white wines. A long time ago, the wine industry figured out that such aesthetic problems could be minimized through the use of fining practices. A type of clay called bentonite is the best fining agent for removing proteins in wine before bottling.

Copper casse and iron-induced casse

A higher concentration of copper or iron can lead to haziness, especially in white wines. These deposits usually present as a white or grey flocculate/haze, which is often referred to as casse. Keeping the copper and/or iron concentration low in wine is an effective way of preventing casse. These days, due to the limitations on chemical spraying in the vineyard and the switch from copper or iron pipes to plastic or silicon hoses in the winery the level of these metal elements in wine is low enough that haze is not a concern for most winemakers.

It's worth mentioning that turbidity can often be a combination of the compounds discussed above. For example, proteins have a tendency to combine with polymerized phenolics and copper casse. Wine producers often use 'colloids' (non-crystalline particles in a solution) as an alternative term for visible particles of any origin.

Tears – what makes a wine 'cry'?

Wine tears, also known as wine legs, are only observed once we have poured a wine into a glass. After swirling wine in a glass, some liquid is loosely attached to its inner wall. This thin layer of liquid slowly moves back down the glass. The visual appearance resembles the shape of tears, or legs, and often draws the human eye. Staining is another term used to describe wine tears, but depending on the definition, staining can be slightly different from tears or legs. Some people use staining to refer to the visible pigments in the tears or film of a red wine in the glass. Intensely coloured and high-alcohol Port wines may leave obvious red or purple coloured 'stains' on the inner wall of the wine glass after swirling.

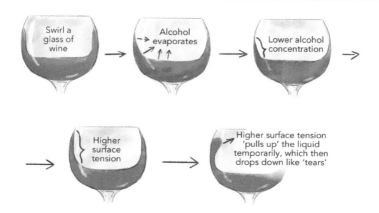

Figure 7: The formation of wine tears

As early as 1855, scientists were able to partially explain the phenomenon of wine tears. It started with the concept of the surface tension of liquid. In water, the molecules cling together more tightly towards the surface, allowing it to resist an external force. For example, a mosquito can land on the surface of water due to the surface tension. The main type of alcohol in wine is ethanol, which has a much lower surface tension than water. At the edge of a wine's surface, where the wine meets the glass, the ethanol content is lower due to more evaporation, resulting in a higher surface tension. Across the rest of the wine's surface, the ethanol content is richer and therefore the surface tension is lower. As a result, the rim or edge with its high surface tension drags up the liquid from below and this flow eventually accumulates into beads or droplets left on the wine glass, which then roll back down, resembling tears. Based on this mechanism, the higher the ethanol content in wine, the greater the discrepancy between the surface tension at the rim and the surface tension in the liquid below. As a result, wines with higher alcohol content have thicker legs.

There are other factors that influence the appearance of tears. It has been shown that the tears of wine are the result of both surface tension and temperature gradients. Additionally, if the sugar content in a wine is very high, such as in a Pedro Ximénez Sherry or Tokaji Eszencia, which can have more than 450 grams per litre of residual sugar, there will be big tears in the glass simply due to the thickness of the liquid.

Bubbles

The sparkling water industry is worth about US$30 billion and its global market grows every year. This strong sparkling drinks economy has facilitated research into and understanding of the science behind the bubble. This section focuses on the appearance of bubbles in wine, while the mouthfeel or texture of the bubbles on the palate will be discussed in Chapter 7. Let's look at the formation of bubbles step-by-step:

Step 1: creating pressure in the bottle

Bubbles in wine originate from an internal or external source of carbon dioxide gas (CO_2) during winemaking or storage. In some instances the alcoholic fermentation generates carbon dioxide and in others the wines can be carbonated by feeding in carbon dioxide gas. In either case, the process happens in a sealed vessel, such as a glass bottle or a stainless-steel tank; the carbon dioxide gas cannot escape and thus dissolves in the liquid. It has been found that the solubility of dissolved carbon dioxide molecules is a function of both temperature and wine composition: the lower the temperature, the higher the solubility (more carbon dioxide dissolved in the liquid). Carbon dioxide is much more soluble in wine than in pure water due to both ethanol and some of the other chemicals in wine.

As carbon dioxide progressively dissolves into the liquid, the pressure in the sealed container increases. Most sparkling wines possess around six-atmospheres of pressure in the newly packaged bottle. But the bottle is not perfectly sealed, so the pressure can be lowered as time goes by. If anyone is lucky enough to taste a sparkling wine aged for decades, they may notice little pressure left in the bottle and the wine will appear only slightly fizzy when poured.

Step 2: creating bubbles in the glass

Without agitation (such as shaking) an unopened bottle of sparkling wine will not display any bubbles. The bubbles form when we release the pressure from within the bottle (by uncorking it, for example).

In a sealed sparkling wine bottle, there is an equilibrium between the dissolved gas in the wine and the vapour phase in the headspace under the cork. The moment the bottle is opened, the pressure in the vapour phase drops and the wine meets the atmosphere. To establish a new equilibrium of pressure between the wine and the ambient conditions, the sparkling wine starts to degas from within.

But not all the released pressure becomes bubbles. When a sparkling wine is poured into a wine glass, there are two mechanisms for carbon dioxide gas to escape. Most carbon dioxide molecules diffuse directly through the surface of the liquid and this process is not visible. The other mechanism is the visible bubbling in the glass. Only about 20 per cent of the dissolved carbon dioxide molecules in sparkling wines escape in this way but they become an important sensory attribute.

To escape, the dissolved carbon dioxide gas changes from the liquid phase to the gas phase. This transformation to a new thermodynamic phase (just like when liquid water becomes solid ice) is called nucleation. To initiate the nucleation, there is an energy barrier to overcome, meaning the initiation of bubble formation requires a certain amount of energy. Think of it as a kind of 'push' needed to get the process started. In the meantime, the surface area of the nucleation site, the area where these bubbles begin to form, is a factor that influences the energy required to form the bubbles. A smaller surface area for the formation of bubbles results in less energy required for gas to form. When we have more unevenness on the inner wall of a wine glass, the higher level of roughness and imperfections acts as more space for the bubbles to form.

This is great news, because there are a large number of microscopic impurities in wine and uneven surfaces on the insides of a glass. They are perfect nucleation sites, reducing the energy barrier tremendously and thus generating bubbles. A more vivid term for a nucleation site is 'gas pocket'. Thanks to these gas pockets, consumers can enjoy a repeated production of bubbles, rising to form elegant bubble trains and bursting at the surface.

The mechanisms of bubble formation in sparkling wines lead to some observations in daily life and certain tricks in commercial settings. If a wine glass is not properly cleaned and polished, the debris will act as a gas pocket to generate more bubbles. This can be counterintuitive to the perception of consumers, as a dirty wine glass (a negative attribute) makes sparkling wine more bubbly (a positive one). Some sommeliers deliberately make tiny scratches inside the glasses used for serving sparkling wines (thus creating additional gas pockets) as a way to make the drink more bubbly during and after pouring. In fact, some branded glassware for sparkling wines has a dent at the bottom of the inner wall to facilitate bubble production.

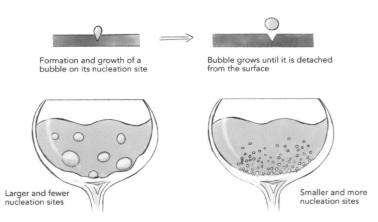

Formation and growth of a bubble on its nucleation site

Bubble grows until it is detached from the surface

Larger and fewer nucleation sites

Smaller and more nucleation sites

Figure 8: The formation of wine bubbles. In a glass where nucleation sites are smaller and greater in number bubbles form more easily.

What affects the appearance of bubbles?

The most direct visual property of bubbles is their size. Bubbles of small size tend to attract people as they give a more refined impression, both visually and, once consumed, through the resulting fizzing sensation (see more discussions on the mouthfeel of bubbles in Chapter 7). Therefore, factors influencing bubble size are worth our consideration.

Bottle pressure

Numerous factors can influence the size of the bubbles in wine. The first and foremost factor is the pressure in the bottle. The loss of pressure (and resultant reduction in carbon dioxide content) during the ageing process of sparkling wines will reduce the size of bubbles. Therefore, old sparkling wines show smaller bubbles in the glass. Some very high-quality sparkling wines can show an even greater level of complexity after long ageing. Hence, people tend to associate the tiny size of bubbles with high quality. But be aware that ageing is not the only contributor to smaller bubbles: cork failure and poor storage conditions can also cause a loss of pressure in wine.

Gravity

Gravity is an important influence on the size of bubbles, even though this may be hard to observe in daily life. If a glass of sparkling wine were to be poured on the moon, the average bubble diameter would increase by 50 per cent due to the much lower gravity on the moon than on Earth. One Champagne producer is testing a piston-activated Champagne bottle at zero-gravity conditions. If the experiments are successful, this particular Champagne label will be enjoyed by some astronauts in the near future. It seems like a lot of effort to go to considering most of us will never get the chance to enjoy a zero-gravity Champagne.

Atmospheric pressure

Atmospheric pressure can influence the size of bubbles. It is something of a tradition for those who have climbed up to the top of Mount Everest to pour a glass of Champagne in celebration of their achievement. They may discover that the tasting experience is quite different when compared with a lower elevation. The atmospheric pressure at the peak of the mountain is less than one-third of the pressure at sea level. In those conditions, the average bubble diameter can theoretically increase by 55 per cent. Therefore, when Champagne is poured at the top of Mount Everest, the wine appears to be significantly more bubbly and very foamy, with much larger bubbles.

Travel distance

The visual effect can differ due to the distance travelled by the bubbles. When a bottle of sparkling wine is poured into a shallow glass or container, such as in a flat, saucer-shaped Champagne coupe, the bubbles barely have any distance to travel from the point of nucleation to the surface of the wine. This means that they float to the top and burst on the surface almost immediately, giving the consumer no time to observe them. When the wine fills a deep container such as a long flute-shaped glass (informally called the Champagne glass), the long travel distance from start point to surface allows the observation of rising bubbles, which can be fascinating to watch.

Temperature

As mentioned earlier, lower temperatures lead to increased carbon dioxide gas solubility. Therefore, if a sparkling wine is chilled to a very low temperature, the carbon dioxide is not easily lost to the air. This slow release of gas will result in a long-lasting bubbly effect. So if you want to observe the bubbles for a longer period, it's important to chill your sparkling wine well. A side note here is that, due to the decrease in pressure at lower temperatures, chilling the sparkling wine bottle also makes it safer to open.

Chemical composition

Lastly, the chemical composition can influence the appearance of bubbles. We all know that the bubbles in sparkling water, beer and wine look different. Compounds such as sugar as well as amino acids, proteins and polysaccharides (released during yeast autolysis, which will be discussed in Chapter 4) in certain sparkling wines will determine their density or viscosity, which subsequently impact the shape of the bubbles. But research is still underway to determine exactly how these compounds help shape the visual effects of bubbles.

Preconceptions and sensory appreciation

The real-life assessment and enjoyment of wine is always a combination of the extrinsic and the intrinsic factors of a product. The intrinsic properties include the look, smell and taste of a wine without any external information. Extrinsic aspects are cues like price, label design, critic scores, brand image and origin. Having such information in our minds can significantly influence our judgement regarding the quality of a wine and affect whether we like it or not. We call this effect upon our judgement preconception. Imagine that we were to present people with a roasted chicken dyed with entirely edible, but blue-coloured, dye. People would find it very strange and might well lose their appetite.

Many studies have proved that taste perceptions are modified toward what one expects, and consumer belief in value versus price dominates the quality assessment of a wine. This is why in almost every sensory experiment, people interact with the samples 'blind', meaning the participants are not aware of the full product information. If we know exactly what we see, taste, smell, hear or touch, preconceptions will develop in our brains and our responses upon interacting with the samples can be significantly modified. (This is also why this book places less weight on subjects such as wine quality, food and wine pairing and consumer preferences, all of which are heavily influenced by extrinsic factors.)

About a century ago, the field of psychology had already taken interest in how past experiences influence the different interpretations of the same image. In terms of food products, numerous studies have proven the effects of food colour on perceived flavours. For example, in an experiment on M&Ms with the same flavour makeup but of two different colours, participants rated brown-coloured M&Ms as being significantly more chocolatey in flavour than green M&Ms. In wine, several studies have explored how the colour of wine can influence the perceived aromas. For instance, a panel of more than 50 people smelled a white wine that had been artificially coloured red with an odourless dye. Without knowing that the original colour of the wine was actually white, the participants described it using aroma descriptors usually associated with red wine.

Preconceptions and solids in wine

Consumer responses towards cloudiness in drinks are highly variable. A study focused on cloudy apple juice grouped consumers by different market segments. One segment liked the turbidity of the juice while another segment disliked how cloudy the juice appeared to be. Another study of apple juice showed that turbidity is perceived as a significant attribute by the participants, but consumer preferences regarding this attribute varied a lot and were influenced by other sensory properties such as the sweetness and acidity of the juice. The exact psychological reasons behind the differing preferences were not clear. It is likely, however, that to certain participants the cloudiness may have equated to 'natural' and 'wholesome', while for others, especially those used to drinking clear apple juice, it may have indicated rusticity.

Most commercial beverage products are clear rather than hazy, as experience causes us to conflate turbidity in liquid with dirtiness and imply it may be unsafe to drink.

However, visible solids can be desirable attributes in a wine if people have a certain level of knowledge or particular preconceptions. With education, consumers understand that many red wines can throw red-coloured sediments after a long period of ageing. Those sediments consist of polymerized anthocyanins and tannins as explained earlier. Experienced tasters even accept heavy sediments when opening certain aged red wines since those sediments become an indicator of an aged product, rather than something undesirable. Moreover, some sediments are associated with the authenticity of specific wines. For example, knowledgeable tasters would question the legitimacy of a clear vintage Port wine without any solids in the bottle, because by law vintage Port must be unfiltered during production.

Other types of wine are expected to have sediments, without having to go through ageing. One example is pét-nat (pétillant naturel) wines, which are sparkling wines made by the so-called 'ancient method'. Due to minimal intervention during production, many pét-nat wines go through little or no filtration, leading to visible dead yeast cells (lees) in the bottle. Such a low level of 'touch-ups' in winemaking is marketed to deliver a sense of purity, romance and respect for nature. As a result, the cloudiness of a pét-nat wine can be perceived as very natural and therefore positive, especially in the eyes of natural wine fans.

Preconceptions and tears in wine

In the early days, when most wines from the cool northern European wine regions struggled to achieve decent alcohol levels, many wines did not show an obvious tearing phenomenon in wine glasses. Back then, higher alcohol content in wine was more likely an indication of riper grapes and better quality. As higher alcohol contributes to more tears (as explained earlier), in the past, a wine with thick legs would have been more likely to be associated with higher wine quality.

Today, the majority of wines display tears after the glass is swirled. That is why there are few studies addressing the sensorial side of wine tears. In a sensory experiment that looks at the influence of alcohol on the sensory perception of red wines, the tears were perceived quite similarly among the samples and thus were not treated as a significant sensory attribute. Anecdotally, consumers can gain a more positive impression of a wine if they note it has thick legs. People in the wine trade often talk about the tears, as they are curious about the heavy staining in association with particular wine styles. For example, fortified wines always have obvious tears, partly due to the high alcohol content. However, how the thickness of legs influences consumer appreciation of wine has not been researched.

Preconceptions and bubbles in wine

Although this is not a book about business, it is worth giving some background information on how much money bubbles can generate. The global Champagne market size was worth around US$6 billion in 2021 and the economy of Champagne and most other sparkling wines is much more powerful than that of other categories of wine. It's not just sparkling wines that draw consumers; the global beer market size was valued at around US$681 billion in 2021 and (as mentioned previously) in the same year the global sparkling water market was valued at around US$32 billion.

Bubbles appeal to people in many ways. As well as the visual experience, the sensory perception of bubbles involves our hearing. Experiments have been conducted to test the perception of sparkling water using auditory cues. The samples were evaluated as being more carbonated the closer to the ear they were held. The celebratory nature of sparkling wines would not be complete without the sound of the 'pop!' as the cork comes out. By listening to the bubbles burst in the glass, another dimension of sensory experience is born.

Many terms are used to describe the visual properties of bubbles: large, small, fine, frothy, creamy, foamy, spumante, fizzy, frizzante, pétillant, etc. The excitement that bubbles bring to people is obvious. An Italian study explored how the appearance of foam in beer influences consumer perceptions. The results of the study showed that beers with a medium level of foam at pouring were the most liked for their visual appearance among consumers.*

Final thoughts

This chapter laid out the chemistry and physics of solids, tears and bubbles observed in certain wines. The sensory section on our preconceptions includes psychological elements, demonstrating that wine is an interdisciplinary subject!

I would like to finish this chapter by using another interesting experiment to showcase how influential those extrinsic cues and preconceptions can be when tasting wines. In one study, a segment of the participants received the positive information that 'You will taste a wine that received 92 out of 100 points from Parker', but another segment was told 'You will taste a wine that received 72 out of 100 points from Parker'.† Of course, all the participants were tasting the same wine. After being given this information, participants tasted the corresponding wine and gave their own ratings. Not surprisingly, those who gave the most positive sentiments were those who were presented with the positive message, and they liked the wine much more than their peers who received negative information.

---

* Similar studies have been conducted on sparkling wines, but the results tend to be proprietary to the producers and not publicly accessible.
† At the time of the study Robert Parker was the most influential wine critic in the world. His rating of 92 points suggests an outstanding wine, while 72 points means a wine with poor to average quality.

PART TWO
Palate

'Dis-moi ce que tu manges, je te dirai ce que tu es.'
(Tell me what you eat and I will tell you who you are)

Jean Anthelme Brillat-Savarin, *Physiology of Taste*

Our palates are the last checkpoint for the toxicity of food before we make a decision on consuming or rejecting it. Prior to the understanding of any chemistry or microbiology that helps ensure food safety, our tongue was (and still is) crucial to survival. Like most other animals, we are reluctant to swallow anything that tastes unpleasant unless it is guaranteed to be safe to consume. However, generations of experience, along with the development of science and technology, have greatly reduced our concern about safety while eating and drinking. Gradually, we became accustomed to – and even started to embrace – certain palate sensations that are supposed to be signs of danger. For instance, coffee can taste incredibly bitter to some people, and bitterness on the palate can be an indication that a substance is poisonous. However, when people figured out that moderate consumption of coffee does no harm to our bodies, the bitterness became the signature taste of coffee, one that we can appreciate or even get hooked on. Our sense of taste has developed beyond the basic signals it originally conveyed and, as a result, the human palate is not as well versed in detecting danger as our vision, which serves a more important role in the modern world (your taste is not going to alert you to an oncoming car as you cross the road, for example).

Like any other sense, perceptions via our palate involve many complex mechanisms. Scientists have only begun to understand the palate to any significant degree in recent decades. In some cases, new findings in taste sensations can completely overrule theories from the past. A famous example of this is the misconception of the tongue map, which has been proved in recent decades to be incorrect (see p. 62). Even today, a few resources, including certain textbooks, still cite this erroneous theory of palate perception. You will see in later chapters that the tongue map was interpreted completely wrongly by some researchers even prior to its being debunked.

The aim of this part of the book is to present the most up-to-date understanding of human palate perception and explain how the components in wine lead to varied sensations in our mouths. Those sensations are mainly grouped into two categories: the five basic tastes and the textural perceptions.

Chapter 4 begins with an introduction to the five basic tastes as the fundamental perceptions via the tongue. Our taste buds are explored and the mechanisms behind taste perceptions are explained. Then the chapter illustrates two fundamental tastes – sweet and umami – as the tastes found in certain wines. The key chemical compounds, such as sugar and glutamate, responsible for these tastes are explored.

Chapter 5 continues to address basic tastes by focusing on the sour and salty sensations. Acidity, which is one of the essential components in wine and is responsible for the sour sensation, is discussed in detail. We explore hypotheses on the chemical origin of the perception of saltiness in certain wines and discuss the ambiguous concept of perceived minerality in wine. The next part of the chapter addresses a more complicated sensory effect produced by a mixture of tastes. The concept of taste adaptation is briefly addressed to reveal the common pitfalls when thinking about food and wine pairing.

In Chapter 6 we round off our discussion of the five basic tastes by looking at bitterness. Phenolics and alcohols are explored in terms of their contribution to bitterness in wine. The chapter also gives an overview of the small amount of recent evidence on the perception of fat as the potential sixth taste. The rest of the chapter dwells on the importance of genetic diversity to showcase how our genetic differences largely determine the presence and intensity of any taste.

Chapter 7 looks at the other perceptions in the mouth when we drink wine. These are the tactile sensations including astringency, viscosity, irritation and temperature. Chemical compounds like phenolics and alcohols are involved once again. Last but not least, the concept of acquired taste is raised to remind us of extrinsic factors such as lifestyle and dining experience that influence our response to any type of palate sensation at a given moment and our judgement of a wine's quality.

4
Sweet sensation and yummy umami

Unlike preceding chapters, this chapter addresses the sensorial terroir prior to the chemical terroir. This is because understanding the basics of our palate is crucial before we explore each type of taste and mouthfeel. It is also important to clarify some misconceptions which still confuse people. After looking at the basics of taste, we'll introduce the chemical compounds responsible for the sweet and umami tastes.

Common taste misconceptions

Let's start with the story of the famous 'tongue map'. It is an illustration that maps our tongue into four areas. Each area is labelled with a specific taste to indicate which section of our tongue is responsible for each taste. For example, according to the tongue map, a section in the front of the tongue accounts for the sweet perception, sour and salty tastes are detected on the sides, and the back of the tongue is responsible for the bitter taste. However, the tongue map is totally wrong.

The tongue map has become one of the classic case studies in how data and information can be interpreted incorrectly, as well as how such misconceptions can be passed down without the original source being examined. The story began with a research paper by a German scientist called Hanig, back in 1901. The paper was the result of a study that concluded that various sections of the tongue are able to perceive all tastes, but there were small differences in sensitivities towards certain tastes among participants in the study. There is no indication of the concept of a tongue map in this paper, other than a sketch of the tongue along with some drawings of data summaries. However, there were many missing or ambiguous aspects in the presentation of data and results in the paper. For example, there was no error bar in the chart.* There was no information regarding the number of participants or statistical confidence level, which should be clearly presented in any research results. It is also worth noting that saltiness was not plotted in the chart because all parts of the tongue perceived saltiness with similar sensitivity during the experiment.

In 1942 the study was reviewed by a Harvard psychologist called E. G. Boring. He resketched the data results in a book called *Sensation and Perception in the History of Experimental Psychology*. Boring did not draw the tongue map, but the way he redesigned the data plotting and chart clearly demonstrated that each section of the tongue had large differences in its sensitivity to different tastes. Such interpretation had no correlation with the results presented in the 1901 paper. In fact, the data in the 1901 paper should not have been used for any serious scientific conclusions in the first place, due to the ambiguities stated previously. These ambiguities carried over to the chart resketched in the book, so this too had missing error bars and no stated sample size. Moreover, remember the salty taste was not in the chart in the 1901 paper? This taste was somehow included in Boring's chart. He is not entirely to blame, though: other researchers have helped shape the tongue map we know today. In the early 1950s, a biochemistry professor suggested in an article in *Scientific American* that there was enhanced sensitivity to a specific taste in a particular area of the tongue but provided no data or evidence to support such a conclusion.

The purpose of presenting the tongue map story is not only to clarify the misconceptions associated with it, but also to show

---

* Error bars are used to present the variability of data and to indicate the error or uncertainty in a reported measurement.

that we have long been in the dark ages of understanding when it comes to the human senses. Thanks to the development of neuroscience, molecular biology and sensory science, we finally have a more accurate understanding of our tongues and how they interpret taste. Certain research has found different sensitivities towards particular tastes in different sections of our tongue, but the results have little consistency, as age, disease, temperature and many other reasons can modify how each part of the tongue perceives certain tastes. We can safely say that the tongue map as a general guide to taste is totally misleading.

Basic tastes

Taste buds are distributed everywhere on the tongue. A few taste buds are also found on other parts of the mouth such as on the surface of the soft palate. More importantly, each taste bud is a receptor for all the basic tastes. Therefore, if one section of the tongue has a lower concentration of taste buds, that section should be theoretically less sensitive to every taste, rather than being less sensitive to one particular taste, as the tongue map would have us believe. A small number of people may lose a part of their tongue through accident or illness, but the rest of their tongue can still taste all the flavour compounds without discrimination.

There are five basic tastes that can be perceived by the tongue of humans: sweet, umami, sour, salty and bitter. It is hypothesized that we possess the ability to perceive a sixth taste, fat, which will be explored in Chapter 6. Our palate can also perceive tactile sensations such as astringency, chilli heat and viscosity (see Chapter 7), but they are not detected by our taste buds. In each taste bud, there are cells acting as the receptors for any stimuli that can contribute to taste. For example, when a sugar molecule meets a sweet receptor, the nerve cells attached to the receptor send signals to our brain which are interpreted as sweet. The five basic tastes are well-established in the academic world as the exact receptors for each type of taste have been validated and the genes responsible for the expression of some receptors examined.

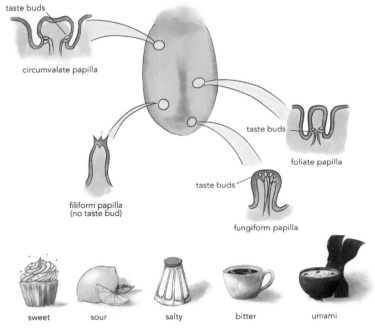

Figure 9: The anatomy of the tiny bumps called papillae which cover our tongue. Inside each papilla (except for the filiform papilla), there are dozens to hundreds of taste buds. Each taste bud can taste all the basic flavours – sweet, umami, sour, salty and bitter.

You may wonder why the tastes presented here are so limited, since a healthy person can perceive thousands of 'tastes' from various cuisines. This is because of another common misconception, or rather a confusion, between the sense of taste and the sense of smell. Many of the so-called 'tastes' in our palate are in fact aromas detected by our nose. Once a given food or beverage is in the mouth, the aromas will travel the short distance from the back of our mouth to the nose for detection. As a result, those aromas are confused with tastes. The distinction between sensations in the nose and in the palate will be explained in Chapter 8. To clarify, all the smells detected by our nose are called aromas, odours or scents in this book. *Taste* only refers to the basic tastes perceived by taste buds in the mouth.

With the foundations of taste in mind, let's explore the chemical and sensorial terroir of sweet and umami.

Sweet

In this section we will be looking at carbohydrates, a group that includes all kinds of sugar compounds. Some of them taste sweet, while others do not. A few chemical properties of sugar compounds are discussed as they are relevant to wine tasting. Later, we will meet other types of compounds which also taste sweet.

Addiction to sweetness

There is almost no one who does not have the ability to detect the sweetness of sugar and almost all babies are born with a strong preference towards sugar. Almost all organic life is naturally born with a 'sweet tooth' because sugar is the main resource for producing energy. Wine, along with all alcoholic beverages, came into existence due to the extreme love of sugar demonstrated by yeasts. Yeasts colonize sugar sources faster than other microorganisms, such as bacteria, and knock out the activities of other microbes when they consume sugar to grow their population. Despite not having taste buds for sensing sweetness, yeasts possess a strong affinity for sugar as their primary energy source and readily convert sugar into alcohol.

Sweetness is one of the most important parameters in distinguishing different wine styles. Wine with a certain level of residual sugar can be perceived as having a sweet sensation on our tongue. A wine with less of the sweet taste of sugar is called a dry wine. These days, dry wines are more popular than sweet wines, in general. This is partly due to the abundance of sugar sources in the world. Since most people can get access to sugar so easily we are now more concerned about type 2 diabetes than starvation. With sugar readily available it becomes less important, taste wise, that the wines we consume have a sweet taste.

In history, however, sweet wines were more highly praised than dry wines. Tokaj in Hungary was the first delimited wine region and the Douro Valley was the first demarcated and regulated wine region in the world. Both regions are focused on the production of sweet wines. The South African Vin de Constance, a sweet wine that was treated as the lifeblood of Napoleon Bonaparte, proved that sweet wines were the nectars of legendary figures. In 1855, one wine was awarded the supreme title 'Premier Cru Supérieur' which represents the highest status in the wine world. That wine was Château d'Yquem, a noble sweet wine that was highly praised from the very beginning. All these examples showcase the prestige of sweet wines in history. This was not only because sweetness gave people pleasure, but also due to the scarcity of sugar sources throughout the European continent until relatively recently. To preserve the sugar source, some Italian wine regions choose to dry their grapes under the sun. The unique wines made from these sun-dried grape berries are referred to as wines made by the 'appassimento' (air-drying) process. Examples include Amarone della Valpolicella, Recioto della Valpolicella and Passito di Pantelleria which are still appreciated by wine lovers today.

Premium sweet wines such as Sauternes, Trockenbeerenauslese, Tokaji Aszú, Eiswein/Icewine and Vin Santo are collected by certain wine enthusiasts. There is a high level of sophistication achieved naturally via the concentration of sugar in the grapes. The grapes are either left to hang on the vines until they shrivel (e.g. Sauternes) or freeze (Icewine), or they are dried to become raisins after harvest. Such production methods not only concentrate the sugar levels but also intensify all the tastes and mouthfeel of the final wines. The results are the absolute intrinsic qualities and the pleasure of appreciating them.

Sugars do not just please consumers of super-premium sweet wines. Inexpensive wines with detectable residual sugars such as white Zinfandel (a pink-coloured wine) have achieved breathtaking sales in volume ever since their creation. The unfermented sugar and the low alcohol levels of those wines turn them into thirst quenchers with a side-effect of merriness. No wonder Mateus rosé, a medium-sweet wine made in Portugal, was one of the most consumed drinks of the American army stationed in Europe during the Cold War.

Sugar has no smell

To be able to taste sweetness (or any other taste), the food must be in liquid form or in a solution. If you were to mop up all the moisture from your tongue using a tissue and place sugar crystals on top of it, you would experience no taste for a short period of time. One of the reasons why we produce saliva in our mouths is to help dissolve solids so we can taste them.

We might have heard someone say of a drink 'That smells sweet!'. While the expression is not necessarily incorrect, it cannot be interpreted as 'That drink must contain a lot of sugar!'. Chemically, pure sugar in a solution does not have any smell. You can prove this by purchasing some sugar with high purity (no flavouring added), dissolving it in water and smelling the sugar-water solution: it should smell no different from a glass of plain water. But in reality, most sweet products contain compounds other than sugar. The aromas often associated with products with a sweet taste are therefore due to the aromas of these other compounds. Those aromas will be explored in the later chapters on the sense of smell, but here is a counter-intuitive wine example. There is a type of fortified wine called oloroso Sherry. When we smell a typical oloroso Sherry, it is filled with toffee- and caramel-like aromas that can definitely give tasters new to the wine the impression that this wine is incredibly sweet. However, the taste of oloroso Sherry in the mouth is often very dry. Other alcoholic drinks, such as many aged brandies and whiskies, smell sweet but taste pretty dry.

The other type of indirect sensorial effect of sugar is also related to our sense of smell. There are some sugar molecules with low concentrations in wine, but they are often bonded with other aroma compounds. Fermentation, ageing and human saliva can break those chemical bonds to release aromas which we then smell. Again, such mechanisms will be discussed in the chapters on aromas.

A chemical breakdown of sugar

Chemically, sugars are carbohydrates that contain the elements of carbon, hydrogen and oxygen. The chemical structures of different types of carbohydrates are not complex at all, and it is exactly because of such chemical simplicity that the majority of living things are able to break down sugars and acquire the energy stored in them very easily.

The simplest carbohydrates are made of a single molecule but other carbohydrates contain more molecules. Let's start with the single molecule first, as there are many kinds of those basic units of sugar. The most common ones are glucose and fructose, which are the most abundant types of sugar in grape berries. Our taste buds perceive glucose and fructose as sweet without the need for further explanation. Such perception is so obvious and intuitive that we only need to dip our tongue into a glass of wine to tell if it's sweet or dry.

But if we taste carefully, different types of sugar give different levels of sweet sensation. Some studies have shown that in water solutions, fructose tastes sweeter than glucose. Sucrose, which combines glucose and fructose molecules together, tastes sweeter than glucose. Sugar cane and sugar beets are abundant in sucrose, so they yield the majority of commercial sugar products. In sparkling wines like Champagne, the final sweetness is adjusted by the addition of the sweet dosage – a 'cocktail' that mainly contains sugar in the form of sucrose (cane sugar). A small amount of dosage (usually less than 12 grams per litre) can completely change the taste of Champagne by making it taste much less acidic.

There are many types of carbohydrates that have more than two units of sugar molecules binding together. Here, we introduce another term for carbohydrate, saccharide, which is derived from the Greek word for sugar. The single unit glucose is a type of monosaccharide; sucrose belongs to the disaccharide group as it contains two units of sugar molecule. In the plant kingdom, carbohydrates can consist of a large number of sugar units, hence the general name polysaccharides for those types of sugars. Starch is perhaps the most familiar polysaccharide, containing between 300 and 1,000 glucose units. Starch is therefore one of the best energy sources in our diets, due to its enormous storage of sugar sub-units. Our taste buds do not perceive polysaccharides as sweet, but the enzymes in our saliva can break down starch into monosaccharides (glucose in this case). Consequently, if you chew some grains of starchy rice, you will slowly begin to taste some sweetness.

Polysaccharides (usually in the specific form of mannoproteins, a combination of the sugar mannose and some proteins) are released into wines from dead yeast cells called lees (discussed in Chapter 3). Especially in sparkling wines that have gone through a long period of ageing on lees, a significant concentration of polysaccharides can be found. While these polysaccharides do not contribute to the sweet taste on the palate, they do contribute to other textural sensations, as we will see in Chapter 7.

Not just sugar

The increasing popularity of sugar substitutes, such as aspartame, used in sugar-free fizzy drinks, has satisfied the sweet taste for consumers worried about consuming too much sugar (although the potential negative health side-effects of these substitutes are still being researched and debated). Food companies conduct a lot of sensory research on how sweet consumers perceive those sugar substitutes to be, in order to ensure the substitute product has a desirable sweetness sensation. In almost all wine-producing countries, sugar substitutes are not allowed in wine, but some other chemical compounds can contribute to a sweet taste. Notably, ethanol, the primary form of alcohol in wine, can be perceived as sweet. But quite a few studies have shown that when tasting pure ethanol solutions at specific low concentrations, some people taste the sweetness, while others perceive different tastes (we will look at this more in subsequent chapters).

Other wine components in lower concentrations have been found to be responsible for the sweet perception in some individuals. Glycerol, mostly a by-product from fermentation, has been shown to have a sweet taste in a water solution. However, it has not yet been proven that glycerol has a significant effect on the sweetness of wine. More facts on the sensorial effect of glycerol and some other compounds in wine are likely to be revealed in the future, as research progresses.

Umami

In this section, we will examine the stealthy sense of umami by showing the history of its discovery. To illustrate the compounds that are responsible for the umami taste, we need to bring some cultural elements into the discussion. The last part discusses the umami taste in wine, in the hope that you will be encouraged to look for it when tasting wine.

The truth was long overdue

To most of us, the taste of umami is perceived by our brains as savoury or 'yummy'. Indeed, the word umami is derived from the Japanese word *umai*, which means delicious. In 1908, Japanese professor Kikunae Ikeda proposed umami as a basic taste experienced by human beings. Basing his work on the taste of kombu seaweed, Ikeda started to delve deep into the chemical compounds that are responsible for the yummy taste in the soup flavoured with kombu. At that time, the culinary world recognized sweet, sour, bitter and salty as the four basic tastes. However, the savoury soup made using kombu has a somewhat distinct taste that cannot be accounted for by those four tastes alone. Eventually, Professor Ikeda isolated brown crystals of glutamate and saw that its taste was what led to the specific yummy sensation. But the recognition of the umami taste was long overdue.

Many cultures and countries have been producing foods that are rich in umami taste for hundreds or even thousands of years. The list of umami-rich produce is seemingly infinite: all kinds of seafood and meats, mushrooms, tomatoes and fermented products like kimchi and soy sauce. Having realized that the umami taste is everywhere in our diets, you may be wondering why it took so long for the scientific field to identify such an important component of taste perception. The answer is that most cuisines and fresh food combine various tastes and aromas and therefore our palate and brain cannot isolate umami as a singular prominent taste. It is much easier to get almost pure sugar for sweetness, acid for sourness and salt for saltiness.

The other reason why the umami taste hides so well is that, put simply, we like it – umami can therefore be enjoyed in all kinds of dishes, mixing with a vast variety of other tastes and aromas. By contrast, bitter compounds are also blended into food, but the 'pain' caused by bitterness is much more pronounced and only compliments certain tastes; therefore bitter tastes have far fewer places to 'hide' and are very apparent to us.

Asian language explains umami

In Japanese, the corresponding Kanji character for umami is 旨 which comes from the right half of the character 鮨, and 鮨 is a type of fish and an alternative term for sushi (note the left part of 鮨 is 魚, which means fish). In Chinese, the word for umami is 鲜 which consists of 鱼 (which again, means fish) and 羊 (which means lamb). The Japanese and Chinese lesson stops here, but the implication of those Asian characters is the origin of the umami taste: something rich in proteins such as fish or meat. Indeed, the major sources of the umami taste are glutamates, amino acids and peptides which are components of proteins.

The chemical source of umami also explains why humans and animals appreciate this taste – proteins and the units of proteins such as amino acids are essential nutrients for us. We need to consume proteins (whether they are from animals or from plants) to sustain our bones and flesh. From the day we are born we seek the umami taste: as infants it leads us to drink breast milk (or milk in general) to enable us to grow healthily. Our metabolism drives the essential requirement for sugar and proteins and the corresponding desire for sweetness and umami.

Back to the 1908 story, the proposed taste of umami did not come from just anywhere. It was also Professor Kikunae Ikeda who isolated a compound with a high level of purity and then confirmed it had a delightful taste that had not been identified in concrete terms before. The compound is called monosodium glutamate (commonly known as MSG) which not only resembles the absolute taste of umami (except for an extra layer of salty taste) but also became one of the most widely used flavour enhancers in the world. Adding MSG in cooking Asian cuisines is a common practice nowadays. But the question we need to answer here is: does the taste of umami play a role in the world of wine?

The afterlife of yeasts

Umami is not (yet) commonly discussed in the wine trade, but the umami taste does exist in certain wines. The key contributors to the umami taste are lees – the dead yeast cells. After fermentation, winemakers can choose to let the wine stay in contact with the dead yeast mass for weeks, months or years. Over time, the cell wall of yeast becomes less rigid and the components within the dead yeast cells slowly release into the wines. Those compounds include mannoproteins (a type of polysaccharide) as discussed earlier (see p. 70), as well as amino acids, peptides, proteins and nucleic acids which largely contribute to the umami taste.

Lees contact or lees ageing are terms that describe the winemaking choice when it comes to leaving the wine on dead yeast cells. Another term, autolysis, refers to the self-breakdown process of the yeast cell and the release of substances such as mannoproteins, peptides and amino acids from within the cell. The degree of autolysis not only depends on the time spent on lees but is also influenced by temperature, pH (the acidity of the environment), alcohol and pressure, amongst other factors. Traditional method sparkling wines like Champagne often go through the deepest autolysis to gain umami tastes and some aromas associated with baking. Therefore, the refreshing acidity in those sparkling wines is balanced by savoury tastes as well as by the sweet dosage. For example, the brut nature type of Champagne has minimal residual sugar (between 0 and 3 grams per litre) but many lovers of this style of Champagne still perceive the high acids as well-integrated without being too dry or too sharp. This is because brut nature Champagne often spends a very long time on the lees, thereby acquiring more umami taste compounds (again, those polysaccharides of mannoproteins, amino acids, etc.) to balance out the acidity.

Umami is one of the predominant tastes in the Japanese alcoholic beverage called sake. Generally speaking, there are seven times more amino acids in sake than in most wines. But even if the umami is not obvious in a wine, such perception can be one of the key attributes that attracts consumers. Given the current trend for lees influence in winemaking, especially in white wines, the role of umami is likely to capture the attention of both researchers and the wine industry. Like sweetness, umami is simply a pleasant taste.

Final thoughts

Taste is a fundamental function in the proper growth and health of human beings. Yet, the biology and mechanism of taste are so intricate that so far we have done little more than scratch the surface in understanding this sense. Misconceptions such as the tongue map have been preached for some decades, but fortunately, they didn't stop us from eventually gaining a better understanding of the sense of taste. Remember that science is always evolving; it would be perfectly normal if some of the scientific findings on taste as reviewed in this book were overruled through experiments and studies in the future.

So far, scientists have found only five (or six) tastes that are detectable by the tongue, but they are sufficient in assisting the basic needs of metabolism. This chapter explained the most welcomed tastes of sweet and umami, which awaken our desire for carbohydrates and proteins, two of the three essential macronutrients constantly used by our bodies. In the next chapter, we will examine the tastes of sour and salty, which reflect the less hungry side of our appetite.

5
Sour grapes and mysterious saltiness

Reactions from new-born babies

Studies have shown that the two tastes sour and salty are less favourable to new-born babies than the tastes of sweet and umami. However, we can develop an appreciation of sour and salty food, as those tastes are related to some essential nutritional elements, such as vitamins and minerals. This chapter explains the wine chemistry of acidity and salt which correspond to the 'less favourable' tastes of sour and salty respectively.

Food and beverages are almost always a combination of tastes: sweet colas come with adequate sourness; umami meat is seasoned with salt. Inevitably, the tastes interact on our tongue and deliver complex sensory information to our brain. Later in this chapter, we will explain a few mechanisms or outcomes of taste interactions, as well as pointing out some aspects of food and wine pairing which could be much more complicated than you think.

Sour

The sour taste in wine comes from organic acids, which are measured and evaluated by different parameters such as titratable acidity and pH. In the wine industry, these measurements can be confusing and their correlations with the sour sensation are not clear. In this section we will look at the most relevant academic information to clarify as much of the uncertainty as possible.

That sour face

Almost every culture understands what a 'sour face' looks like. When too much acidity is perceived by the tongue, our brain reacts in a negative way and our subsequent grimace indicates the unpleasantness we experienced, along with a tightening of the facial muscles to attempt spitting. We naturally detest too much acidity in our food and beverages as part of our protective systems against unripe, toxic and even corrosive objects. But with the right amount of acidity in food and drinks, the tasting experience can be an absolute pleasure. Acidity awakens our appetite and gives a refreshing mouthfeel. Such a pleasant response to sourness is due to the nutrients that we need at lower concentrations, compared to the larger amounts of sugar and proteins that we need. For example, fruits rich in vitamin C (which is itself a type of acid, called ascorbic acid), such as lemons, have abundant acidity and humans around the world have learned to embrace them by adding small amounts of the sour lemon juice to their cuisines to balance out other flavours. In contrast, very few people appreciate the high concentration of acidity that comes from eating the flesh of lemon without any dilution. Our palate has been well-designed to seek just the correct level of mesonutrients and micronutrients.

Acid is the third most abundant component in wine after water and alcohol. A wine can still be a wine without tannins, sugar and some aroma compounds, but a wine must have acidity due to the natural high acidity in the grapes. Apart from containing alcohol, it is the acidity that makes a glass of wine unique. The acids in wine have numerous chemical and sensorial properties. The role of acidity in wine goes far beyond its contribution to sourness.

An 'anatomy' of acids in wine

As mentioned, all acids in grapes and in wines are organic acids, meaning they are weaker than inorganic acids such as the corrosive sulphuric acid, thus making a wine safe to consume. In grape juice, the major types of organic acids are tartaric acid, malic acid and ascorbic acid. During winemaking, yeasts and bacteria can produce and modify certain organic acids to yield succinic acid, lactic acid and acetic acid. Other types of acids such as sorbic acid can be added to wines in small concentrations for the purpose of stabilization.

One of the special features of grapes is the incredibly high concentration of tartaric acid, which usually accounts for more than 70 per cent of the total acids in grapes. What's even more special is that tartaric acid cannot be broken down by most microorganisms. Thanks to the great stability of this type of acid, archaeologists and chemists have found evidence of winemaking thousands of years ago through the remaining tartaric acid in amphoras discovered at ancient archaeological sites. Also, thanks to alcohol and tartaric acid, since ancient times, wine as a stable agricultural product has become one of the key driving forces of human civilization. The stability of food products like wine allows transportation and storage, which founded the basics of agriculture, trade, migration and colonization in the early days of civilization.

Other types of acids found in wine include malic acid, citric acid and ascorbic acid, whose concentrations vary depending on the surrounding environment and the condition of the grapes. Malic acid, for instance, can drop in concentration significantly under hot conditions or under disease pressure. What's more relevant to winemakers is that during winemaking malic acid can be converted by bacteria to the seemingly softer-tasting lactic acid. Yeasts, on the other hand, produce succinic acid during fermentation. Some scientists and winemakers believe that succinic acids have significant impact on the sensorial outcome of a wine, but the sensory properties of this type of acid are not well-studied. Not long ago, the yeast species *Lachancea thermotolerans* was found to be able to produce lactic acid as a fermentation by-product. Some beer and wine producers therefore use these yeasts to increase the acidity. As for acetic acid, we will see in Chapter 12 how its low concentration in wine means it doesn't contribute to much sourness, but its volatile nature contributes to some smell.

Chemistry 101: pH and total acidity

Chemists devised the most useful way of measuring acidity more than a hundred years ago. One of the most widely used parameters for gauging acidity is the famous pH scale, which gives an intuitive idea of how acidic or alkaline a substance is. For example, pure water is qualified as 'neutral' because it has a pH value of 7.0, which is neither acidic nor alkaline. The key points to keep in mind are:

* The lower the pH, the more acidic a solution is.
* The higher the pH, the more alkaline a solution is.

Any solution with a pH value below 7.0 is considered acidic and those solutions with pH values above 7.0 are alkaline.

To understand the significance of pH in wine, the chemistry lesson goes a bit further, toward the actual definition of pH: the decimal logarithm of the reciprocal of the hydrogen ion activity. If you are not familiar with higher-level chemistry and mathematics you may not understand this definition but don't worry, there is only one important point to keep in mind here: pH is a measurement of the hydrogen ion (H^+) concentration in a solution.

To chemists, those ions are significant, as each ion readily possesses more energy or reactivity than molecules without a positive or negative charge. For example, concentrated sulphuric acid solution has a pool of an extremely high level of hydrogen ions which contributes to its corrosive and destructive nature. In comparison, lemon juice contains a much lower concentration of hydrogen ions, although our tongue still feels the 'attack' from these ions as sourness.

There are different categories of acid; some are classed as strong and others as weak. What this relates to is the proportion of their constituent molecules that release their hydrogen as hydrogen ions. Some strong acids, like the sulphuric acids described above, release all of their hydrogen elements as hydrogen ions. Organic acids in wine are the so-called weak acids, which only dissociate

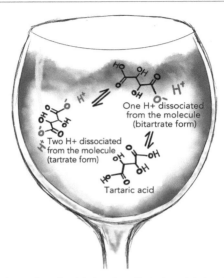

One H+ dissociated
from the molecule
(bitartrate form)

Two H+ dissociated
from the molecule
(tartrate form)

Tartaric acid

Figure 10: The dynamics of acids in wine. Tartaric acids in wine take three forms – tartaric acid, bitartrate and tartrate. The concentration of hydrogen ions (H⁺) dissociated from the acids will determine the pH of the wine.

part of themselves and release a proportion of hydrogen ions from the acid molecules. This is why for most food and beverages, measuring the total concentration of acids does not reflect the concentration of readily released hydrogen ions which is perceived by our taste buds as sour, as the majority of hydrogen ions are still bound up in the structure of the molecule.

In addition to acidity, pH provides winemakers with other crucial information such as the wine's microbiological stability, colour in the case of red wine (more red or more purple as seen in Chapter 1) and the effectiveness of sulphur dioxide and enzyme additions. The concentration of hydrogen ions has an influence on all of these factors. The dynamics of wine chemistry are heavily influenced by the level of pH, but the science involved is beyond the scope of this book.

Total acidity, also called titratable acidity (the two are slightly different when performing chemical analysis but equivalent for our purposes), is the sum of all the available hydrogen ions in a solution, but the hydrogen ions here combine the free form (already dissociated in the solution) and the bound form (not yet released as free hydrogen ions). In other words, pH can be viewed as the amount of hydrogen ions readily there in a solution, whereas total acidity also includes the H^+ ions which can potentially be released. A wine can have a high level of total acidity, but with a low level of free hydrogen ions (higher pH) due to factors like high potassium in the solution (think of potassium as having the ability to take the place of hydrogen ions in grape juice and in wines). To get a more comprehensive idea of the acidity of a wine, winemakers nowadays examine both the pH and the total acidity. But can chemical parameters like pH and total acidity explain the sour, acidic perception on the palate? The answer is yes, but we will need to delve a bit deeper into the chemistry and sensory mechanisms to understand this.

Sensory 101: pH and total acidity continued

First of all, it should be noted that most foods and beverages do not have a linear relationship between their acidity and how sour they taste. This is because of many other variables (you might say 'noise') that can hide sourness. For example, many classic and premium sweet wines such as Sauternes, Tokaji Aszú, Icewine/Eiswein, Beerenauslese Riesling, Coteaux du Layon, etc. are supported by low pH and high levels of total acidity. However, because of the sweetness, many people don't perceive them as sour while tasting. Such taste interactions between sweetness and sourness will be explored later in this chapter.

During the winemaking process, a winemaker can encourage or block the conversion from malic acid in wine into lactic acid. Quite often, we hear that malic acid has a crisp taste – such as we experience when biting into an apple – whereas lactic acid tastes softer, as we can perceive when eating yogurt. This indicates that each type of organic acid may contribute to a different kind of sour sensation. Yet, when researchers conducted relevant experimentation, there was no clear answer regarding whether some types of acid taste more sour than others. In one study, lactic acid was found to be significantly more sour than citric acid, while other studies showed that citric acid contributed to a higher intensity of sour taste. Some studies also saw that different participants respond to the same acid solution quite differently from one another. Therefore, in practice, when winemakers create their wines, it makes much more sense to focus on pH and total acidity, considering these data are much easier to obtain and to ensure accurately.

Early studies have already shown that both pH and titratable acidity appear to be important in determining the sensory response to sourness. In order to understand how the two measurements reflect the sour taste, we need to consider two types of perception of sourness:

- How strong the sour sensation is at a given moment.
- How long the sour sensation lasts on the tongue.

The former is easy to explain because it is the contact between our taste buds and the hydrogen ions that gives the sour perception. As pH is the direct reflection of how many free hydrogen ions there are in a solution, low pH is directly correlated with the strength of sourness the moment when the hydrogen ions touch our tongue. Therefore, in theory, a low-pH wine tastes intensely sour if we focus on the second when it enters our palate. The second point, however, is most relevant to how we experience tastes.

In reality, the organic acids in food and wine stay in the mouth for a while even if we swallow (or spit) quickly. This is why the lasting sensation of sourness is more relevant, yet it is more complicated to explain. Let's review what we saw earlier in Chemistry 101 of pH and total acidity. The organic acids in wine are weak acids which only release some hydrogen ions. In other words, without disturbing the chemistry in a glass of wine, there is a state of balance between the bound hydrogen ions and the free hydrogen ions. But once we take wine into our mouths, our saliva, which has a higher pH than the wine, starts to neutralize the hydrogen ions. As the level of free hydrogen ions decreases, more hydrogen ions start to detach from the bound form until all the available hydrogen ions in the acid molecules are depleted. If acids were people and hydrogen ions were currency, the strong acids would tend to spend all of their money upfront, whereas the organic acids always spend much less in one go, so their cash flow lasts longer. During wine tasting, saliva is constantly produced to neutralize the hydrogen ion attack on the palate. The more hydrogen ions reach our palate, the more saliva we produce. This is why many wine professionals suggest the best way of evaluating acidity is to see how long a wine makes your mouth water.

If you compare how long the sour sensation lasts among different wines, the durations are certainly different. There are two factors that determine such sensorial differences. One is the concentration of total acidity, which in light of the explanation above we now understand, since a higher level of organic acids in a solution means a larger pool of molecules that can donate more hydrogen ions. The other factor is related to pH, but this time, pH must be associated together with another value called the pKa. At the same temperature, each type of acid has its own 'acid dissociation constant', the pKa value. Understanding the concept of pKa can be intimidating, but for our purposes we just need to understand that whenever a solution's pH is equal to the pKa value, that particular solution has its strongest buffer capacity, meaning it is highly resistant to releasing hydrogen ions too quickly.

For example, tartaric acid has two pKa values as each molecule can release two hydrogen ions. At room temperature (25°C), one of the values is pKa = 2.98 and the other is pKa = 4.34 (see more discussion on temperature below). Since the pH values of most wines are in the range of 3.0 to 4.0, we can assume that if the pH of a wine is around 3.0 (a very acidic wine) or about 4.3 (a wine with very low acidity), either pH level gives the wine a strong buffer capacity, leading to a long-lasting saliva-producing effect in our mouths. But keep in mind that there are other organic acids in wine that can shift the pKa values. For example, lactic acid has a pKa value of 3.86, which can change the overall pKa value in a wine, depending on the concentration. Due to such complexity, winemakers rarely measure and calculate the exact pKa of a particular wine.

In summary, pH indicates the initial 'attack' of acidity on our palate, while the total acidity and the discrepancy between pH and pKa determine the lasting effect of sourness. Imagine a bottle of wine that had a strong sour sensation from the very beginning, if that sensation lasted for a long time, what would the chemical parameters be like? It would need a pH of 3.0 (or lower), which gives a strong sour impact when the wine enters the palate. It would also require a high concentration of total acidity; around 10 grams per litre (or higher), which further ensures the abundant pool of organic acids to donate the hydrogen ions responsible for the acidic sensation. Furthermore, it would need to have an overall pKa value of 3.0, the same value as the pH (recall that this provides the strongest buffer), which means that the hydrogen ions are released at the slowest pace possible to ensure the lasting sour perception in our mouths. (Champagne often comes close to meeting all these paramenters.) But note that some other factors can alter such chemical makeup. The pKa value(s) of each type of organic acid is dependent on temperature. At room temperature, one of the pKa values of tartaric acid is almost 3.0 (2.98 as mentioned in the previous paragraph), but the value increases as the temperature decreases. This means that when we chill a wine before consumption, the dynamic of pKa values changes. Consequently, the duration of the sour sensation in our palate changes when we enjoy the wine at different temperatures.

Salty

The salty taste in wine is more mysterious than the other flavours we have discussed this far, but has recently become more frequently discussed in the wine trade. Here we look at some evidence and hypotheses regarding the origin of saltiness in wine.

Enhancing life

Human beings naturally discovered the magic of salt in cooking. The majority of dishes in the world are seasoned with salt. Without the salty taste found in many dishes, the food would taste bland. We need salt (sodium chloride) as sodium serves some vital functions in our body such as conducting nerve impulses, regulating the functionality of muscles and maintaining the balance of water and minerals. Like acids, salt is essential to us, but too much of it can be harmful. An excessive intake of salt is correlated with health issues like high blood pressure, heart disease and kidney disease.

Our ancestors were smart enough to take advantage of the toxic effect of salt when used in high concentrations. It was one of the first sanitizers and preservatives, though at the time humans did not understand the mechanisms involved, which are that salt reduces the water content in preserved food and therefore prevents unwanted microbial growth such as moulds. In many cultures, people used generous amounts of salt to preserve food for a much longer period of time. As early as 2700 BCE, the Chinese used salt water as a mouthwash to promote gum health.

Salt is sodium chloride

The chemical makeup of salt is sodium chloride (NaCl). It is easily obtained from sea water, but is not something you would expect to encounter in wine; wine grapes naturally contain extremely low concentrations of sodium chloride. Our taste buds sense the sodium ion (Na^+) in a sodium chloride solution as salty in taste.

In theory, wine as a natural product of fermented grape juice should not have any obvious taste of saltiness. However, there are indeed wines with a perceivable salty sensation, such as the famous fino and manzanilla Sherries. Some people suggest that the salty taste comes from the special status of yeasts called the *flor* that develops during winemaking. The *flor* yeasts release compounds like acetaldehyde which can give a tart taste that is close to saltiness.

Another theory is that the sodium chloride concentration may be higher than normal in these Sherries due to the ambient environment of Jerez – the Sherry production region. During the maturation stage of Sherry, the rooms staging the wine barrels are exposed to the ocean breezes which naturally provide cooling influence and humidity. Consequently, the wind off the sea – carrying salt – is introduced into those Sherries. Since salt is rarely a relevant compound during winemaking, little data are available regarding the final sodium chloride concentration in fino and manzanilla Sherries. However, the ocean-derived salt influence has been observed in many windy coastal regions in the world. On the Greek island of Santorini, the outdoor fermentation tanks can be covered with thin salt layers in a few hours due to windy conditions blowing in salt water from the sea. It would not be surprising if some sodium chloride found its way into the fermenting grape juice to add saltiness.

The 'minerality' debate

In the wine industry, one of the most debatable concepts is the mineral sensation or minerality of wine. The reason for the ongoing discussion is that the concept of minerality cannot be quantified and the definition of the term varies among people.

A mineral is defined as a natural solid with a defined chemical composition and crystalline structure. In the wine industry, though, the term minerality can refer to any sensation in wine that reminds people of earth, rock or any inorganic matter that seemingly has high mineral content. Terms such as 'wet stone', 'flint' and 'chalk' appear frequently when describing a wine with a mineral taste or smell. But it is important to keep in mind that these mineral perceptions are not related or relevant to actual mineral elements in wine. This is because most of the mineral compounds, such as those involving the elements of iron and copper, are in very low concentration in wine. The most abundant ones include potassium, which may contribute to a decrease in the taste of sourness by binding with tartaric acid to form potassium bitartrate, which has no taste, but they are hardly correlated with 'minerality' as described by people, not only because the mineral content in wine is very low, but also due to the fact that the mineral compounds associated with potassium, iron, copper and calcium ions in wine do not contribute to any sensory quality.

The two sensory qualities discussed in this chapter are speculated to have correlations with the experience of mineral sensation. A wine with high acidity or with a salty sensation can be perceived as significantly more mineral than one without those attributes. Beyond the taste on the palate, there are specific smells in wine that are considered mineral-like and Chapter 12 will touch on aromatic minerality. In addition to the complexity and vagueness of the sensory aspects of minerality, research has also demonstrated that a person's culture influences the way in which they perceive and describe the mineral sensation. For better communication, perhaps the description of minerality ought to be more specific, such as 'wet stone-like acidity' or 'a sensation of salty minerality'.* Whether or not such expressions can be understood or shared by two people is another matter.

Taste interactions

As most foods and drinks provide a mixture of different tastes, taste interactions happen all the time in our lives. Lots of the things we eat and drink are about the balance between sweetness and sourness. For example, the levels of sugar and acid can be adjusted to tailor the most delicious lemonade to an individual taste. For about 30 minutes after eating the miracle fruit (*Synsepalum dulcificum*), people lose the ability to taste the sourness of lemon juice and instead can taste only the sweetness of the lemon. The discussions on pH addressed the high acidity behind classic sweet wines such as Sauternes and Tokaji Aszú.

---

* In sensory science, when conducting a type of research called 'descriptive analysis', panellists need to be trained with items that can be found in real life, such as tasting a real piece of wet stone (without swallowing it of course) and perceiving different concentrations of salt solutions, so that all the participants can evaluate the corresponding sensory attributes in wine using those reference standards in consensus.

But people rarely perceive those wines as too acidic since the sour perception is masked by the high sugar content.

Scientifically, we still don't have a comprehensive understanding of each of the five tastes. For example, it is still not clear how the reactions in our brain switch from positive to negative as the intake of salt increases in concentration. Thus, studies on taste interactions are even more complicated due to the multiple factors and variables involved. The good news is that recent research has started to shed light on this subject and point towards promising areas for further research, in order to decode the complexities found thus far in sensory-related studies.

The ionic attack

Of the four tastes that have been discussed so far, sweet and umami are similar to each other in the sense that the compounds responsible for both tastes (sugars and amino acids) are required by our body in large quantities. But as we have seen, we need much lower concentrations of acids and salt for essential metabolism: sour and salty can therefore be treated as another pair – taste qualities that prevent us from consuming excessive amounts of acids and salts. Viewing the four tastes as two dissimilar groups is supported by genetic and neurological studies on taste receptors. Although further validation is required, it is most likely that the receptors for sweet and umami (and also bitter) tastes are special protein structures called the G-protein-coupled receptors. The receptors for sour and salty tastes are ion channels that allow the passage of ions of hydrogen and sodium for further reactions that signal the sense of sour and salty in our brains. Therefore, when organic acids and sodium chloride make contact with our tongue, they are like legions of ions marching through channels in the taste buds. Like most ion channels in cells, they regulate the amount and the transit speed of the ions that attempt to pass through.

Interestingly, the ion channel responsible for sensing sodium ions also allows some hydrogen ions to pass through, meaning the concentration of sodium ions that stimulates the receptor is slightly diluted by the intrusion of hydrogen ions. This mechanism may explain why the addition of lemon juice can make a highly salted dish taste less salty. There is also evidence to show that sodium ions interfere with the bitter taste transduction* on the palate, which helps to explain the reduction of bitterness when tasting salt at the same time.

Most studies into taste interaction have examined the binary taste interactions of two of the same or different types of taste. There are three kinds of effect observed:

- One taste enhances the other.
- One taste suppresses the other.
- The two tastes remain at the same intensity.

Because of the variabilities among the participants and the experimental conditions, such as the concentration and temperature, varied results have been reported. However, thanks to the review work done by diligent researchers, certain patterns were found to remain consistent across a number of studies. It was found that at higher intensity levels:

- The sweet taste and sour taste suppress each other.
- The sweet taste and bitter taste suppress each other.
- A salty taste suppresses bitter taste when present in moderate intensities.
- A bitter taste neither enhances nor suppresses a salty taste when present in moderate intensities.
- At lower intensity levels, the salty taste and sour taste enhance each other.

Last but not least, it is worth noting that people can confuse specific tastes in a mixture. Quite a few experiments have proven that sour, salty and bitter can be easily confused. Such confusion

---

* A process by which a signalling molecule such as a bitter stimulus binds to the taste receptor and changes the receptor protein in some way to make our brains interpret a specific taste.

is heavily influenced by the types of food and drink a person habitually consumes, as well as the language they speak. For example, the taste of citric acid may be identified as bitter or salty by some people. But after receiving training in tasting citric acid against sea salt and distinctly bitter compounds like quinine, those people would recognize citric acid as sour rather than salty or bitter. No wonder so many wine tasting notes get lost in translation.

The reality of food and wine pairing

All organic life is constantly adapting to changes in the environment. If we hold a lusciously sweet wine on the palate, the intensity of the sweet taste will decrease over time. This is called taste adaptation and other senses, such as vision, have the same property. For example, if you were to step into a cold shower, you would feel the shock of the cold to begin with but your skin would become less and less sensitive to the low temperature of the water. In Chapter 10, we will explore odour adaptation in detail. But it is important to bring up taste adaptation here to clarify the misconceptions inherent in the popular idea of food and wine pairing.

It is difficult to delve into the technical aspects of food and wine pairing, because both elements have their complex chemical and sensorial natures. Experienced food and wine professionals have developed some rules that generally work well, however. For example, dessert should pair with wines of equal or higher levels of sweetness, otherwise the wine can taste tart or overtly sour (due to the existing sweetness of the dessert desensitizing the palate to sweet tastes). Such guidelines are important in the restaurant business, especially in terms of avoiding pairings that tend to result in unpleasant tastes, like too much of a sour or bitter sensation in food or in wine.

However, taste adaptation is rarely considered in food and wine pairing. The concept really should be given greater attention, given that taste adaptation plays an important role. This is immediately obvious if you consider the way in which we dine. The food and the wine are never pre-mixed for us to taste the mixture together (except perhaps in very rare specific circumstances). When dining, we alternate eating food and drinking wine: our palate therefore will become accustomed to one taste before we experience the next. For instance, drinking Champagne that is high in acidity before a meal can make a dish dressed with lemon juice taste more flabby and less acidic because the constant intake of Champagne makes us less sensitive to the sour taste. On the other hand, the same dish acidified by lemon juice tastes significantly sour when our palate has not been bathed in wine with high acidity.

The situation becomes more complex and dynamic when people choose to sip some water between food and wine consumption. Moreover, people who eat and drink slowly experience more taste adaptation effects than those who consume at a fast pace. In summary, a meal with food and wine together is more about adaptation and the stimulation of new tastes that happen at different times. For chefs and sommeliers, the safest approach is always to try out the food and wine combination in the way in which it would be consumed by a regular diner, in order to experience how a customer may perceive the dining experience.

Final thoughts

In this chapter, we broke down the chemical matrix of acidity in wine and explained the theoretical linkage between titratable acidity, pH and the 'sour' taste. The concepts of saltiness and minerality in wine were also explored. Hopefully, your brain has retained some aftertaste of what you have read about sour and salty. More importantly, I hope you are now aware of the importance of taste interactions in our everyday consumption of food and drinks. This may help you to enhance your own dining experiences. Rather than using pre-ordained matches between food and wine, I invite you to try out the countless food and wine combinations, as well as adjusting other variables such as changing the order in which you eat and drink the elements of your meal and seeing how the environment, mood and conversations during eating and drinking alter your experience of taste.

6
Some bitter truths about taste

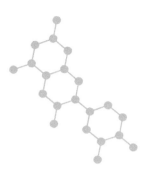

Survival instinct

In terms of food safety and survival, human beings have long relied on the knowledge and behaviours passed down by our ancestors, and because of this both our need and our ability to distinguish poisonous substances for ourselves has gradually decreased. (We have also lost much of our connection with nature, so to some audiences, survival reality TV shows give useful tips on how to find the edible items in nature.) To animals in the wild, the ability to taste bitterness is as important as pursuing sugar and proteins. Many toxic compounds in nature are perceived as bitter or some other unpleasant sort of sensation. The taste buds of animals act as the last checkpoint that can reject poisons relatively easily. Once anything toxic is swallowed, it is hard for it to be ejected or removed from the body. Lots of medicines taste bitter, too. You may have bemoaned the awful sensation of accidentally biting on a paracetamol tablet, for example. Like the salty and sour tastes, the bitter sensation prevents us from consuming too much of those remedies which generally become harmful at high doses. Anyone who has been around a toddler who wants to put everything in their mouth likely has a greater appreciation for these foul-tasting medicines; if the toddler were somehow to snatch up a stray tablet, the bitter taste would be likely to make them quickly spit it out.

In this chapter, both the chemical and the sensorial discussions have a common theme behind them: our genetics. Scientists hypothesize that our basic tastes evolved a few hundred million years ago for survival and reproductive purposes. With regard to our perception of bitter and fatty tastes, the underlying genetic makeup is rather complicated and therefore researchers are still working hard to gain a better understanding of the bigger picture. The accumulated mutations and diversification of the taste-related genes not only make it hard for scientists to decode which gene is doing what, but also result in distinct genetic diversity among the human population in terms of certain tastes. This is especially true when it comes to bitterness: people's perceptions and acceptance of this taste can be vastly different.

Is there bitterness in wine?

Quinine is a bitter compound that has the medicinal property of treating malaria. More than a century ago, quinine was added to many different types of alcoholic drinks to serve as a conventional medical way to prevent and treat diseases. Existing bottles of Sherry from the late nineteenth century or the beginning of the twentieth century will taste bitter because of the quinine added. To counterbalance the bitter taste, those Sherries tended to be made in a sweet style, with residual sugars. Today, the most popular drink that contains quinine is tonic water. No wine today should be allowed to have quinine added as wine laws across the world restrict the types of additives permitted.* However, the taste of bitterness still exists in certain wines. Many studies have proved that the phenolic compounds in wine are the main contributors to bitterness.

In the next chapter, tannins, which are responsible for the astringent mouthfeel, will be discussed in detail. Here we briefly need to address the fact that tannins can also have a bitter sensation. This is especially true for most people if something

---

* Personally, I am also interested in exactly when quinine was made an illegal additive to Sherry: chemistry and sensory science aren't the only areas that make me excited.

sweet is consumed first, and then a dry red wine with lots of tannins is tasted; in this case the wine can taste quite bitter. Research has revealed that, depending on the composition of tannins (tannins are always a combination of multiple phenolic units, as we will see in Chapter 7), the perception of bitterness varies. During the production of red wines, tannins can be extracted from both the seeds and the skins of grapes. Seed tannins tend to taste more bitter than skin tannins. Also, tannins that contain smaller units taste more bitter than larger-sized tannins with multiple units combined together after extended ageing (again, something we will cover in the next chapter).

Other than tannins, which belong to the group of flavonoids amongst all phenolics, there are other phenolic compounds that are categorized as non-flavonoids (discussed in Chapter 2). These non-flavonoids are important in most white wines that don't have a significantly high concentration of tannins extracted from skins and seeds. Due to the low concentration, it seems that non-flavonoids rarely contribute to bitterness in white wines. However, quite a few wine tasting notes may refer to the 'phenolic bitterness' in white wines made from grape varieties such as Pinot Gris, Viognier and Gewürztraminer. Perhaps certain non-flavonoids are responsible for some bitter taste, or it is because those white wines have sufficient levels of bitter tannins extracted from the grape skins and seeds during winemaking, or even because some bitter oils within the grape seeds were released due to a deep level of juice pressing.

Alcohol can also contribute to the bitter taste, but it can also be perceived as sweet or burning at different concentrations by different people. More discussions on the variable tastes of alcohol itself will be found in the next chapter.

As we can see, there are a good many compounds that are perceived as bitter by our taste buds. Over time, our body has evolved multiple genes to perceive signals of potential risks but thanks to the development of food safety, there is no need to lick most food before consumption nowadays.

The sixth taste: fat

Though not a taste encountered in wines, let's take a brief moment to mention the taste of fat here, as an indicator of the ever-changing nature of science. It is possible that this sixth taste will be confirmed soon and become an important consideration in fatty food and wine interactions.

Although the exact mechanism is unknown, the perception of fat by the tongue is widely believed to be the sixth taste perceived by the taste buds. The compounds responsible for the fatty taste are considered to be free fatty acids, but in wine the concentration of such compounds is so low that they should not contribute to its taste. The more important role of fatty acids is to influence the activities of yeasts so that interesting aroma compounds can be produced (in Chapter 8, fatty acids as precursors to aroma compounds will be mentioned). Those non-volatile fatty acids that make contact with our tongue certainly contribute to particular sensations, but much like the umami taste discussed in Chapter 4, we rarely taste them in isolation. Additionally, there are many chemical forms of fatty acids, making the understanding of the related perceptions difficult to comprehend, at least for now.

Genetic diversity

Unlike colour-blindness, which is not uncommon among the male population, it is rare to find people with taste disorders unless they are exposed to harmful chemicals, receiving radiation therapy for cancers, suffer from a brain injury or are ill with diseases such as gastroenteritis, etc. There are three main taste disorders and sets of symptoms:

- Dysgeusia, the symptoms of which mean an individual experiences a foul, rancid or metallic taste that persists in the mouth;
- Hypogeusia, when the ability to taste is reduced significantly;
- Ageusia, which is extremely rare among the human population, and means the inability to taste anything.

Due to the rarity of these taste disorders, in the following discussion on genetic diversity we will focus on the observations of people without severe abnormality in taste, so as to be more generally relevant.

In Chapter 2 we explored the variations in human vision due to genetics. Our genetic diversity is far more pronounced in the realm of taste. Without considering psychological factors (which also play an important role in taste perceptions), the sensorial terroir shows an exciting world of different sensitivities among us towards specific tastes. Interestingly, scientists only began to understand this genetic diversity in taste and in smell in recent decades. Here we will look at some of the important discoveries that are relevant to wine tasting when it comes to variabilities in taste.

'Supertasters', hypotasters and everything in between

The reason I placed quotation marks around the term 'supertasters' is that it is a misleading term. 'Supertasters' refers to a group of people who have significantly higher sensitivity to basic tastes like bitter and sweet, as well as to tactile sensations like the burn from alcohol. But calling these individuals 'supertasters' makes it sound like they possess some kind of superpower in their palate and are somewhat superior to others, which is not true. In fact, such heightened sensitivity often proves to be a drawback. In daily life, 'supertasters', with their far greater sensitivity, tend to be quite picky when it comes to food and often find alcoholic beverages, black coffee and spicy cuisines very unpleasant. They may find something as simple as a sip of coffee, even one filled with sugar and cream, overwhelmingly bitter and dreadful on their palate. Far from being imbued with a superpower, 'supertasters' may not be able to enjoy some of the common culinary pleasures most of us have on a day-to-day basis. A better term for these people is hypertasters, which communicates accurately that those people are simply more sensitive towards tastes.

The opposite of hypertasters are the hypotasters, who tend to be very insensitive to a lot of taste stimuli. Hypotasters can generally tolerate what are perceived as high levels of bitter, spicy and burning sensations by most people. In some cases, hypotasters cannot perceive specific taste compounds at all. For example, the compounds of phenylthiocarbamide (PTC) and propylthiouracil (PROP) have an obvious bitter taste to about 70 per cent of the population of the world. The other 30 per cent do not perceive anything with either of the two compounds in their mouth. There are variations in terms of perceiving all the other four tastes, but bitterness is certainly the sensation that showcases genetic diversity most obviously.

A study examined the correlations between people's sensitivity to PROP and their taste perceptions of red wines. The participants who could not taste PROP at all demonstrated significantly lower intensity ratings for the bitterness, astringency and acidity of the red wines tasted when compared to those who perceived PROP as bitter. It seems that the sensitivity towards one type of taste, especially bitterness, can indicate that a person is sensitive to other tastes. In other words, hypertasters tend to be sensitive to all types of tastes on the palate. This can be explained by looking at the anatomy of the tongue. Studies have proved that hypertasters have a much higher concentration of taste receptors compared to hypotasters.

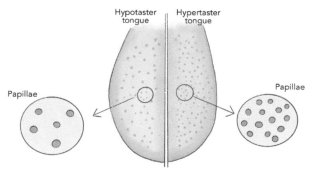

Figure 11: Hypotasters versus hypertasters. Hypertasters are more sensitive to the basic tastes than non-tasters. Correspondingly, the former tend to have a much higher concentration of papillae (which contain taste buds) than the latter.

There is no distinct line that exists between 'normal' tasters and hyper/hypotasters. Most people are neither super-sensitive nor super-insensitive to taste stimuli, but that does not mean that there are no variations among 'normal' tasters. In fact, the variations are huge regarding our sensitivities toward various tastes. As noted, bitterness is the type of taste that can demonstrate our taste sensitivity the most clearly. Among 'normal' tasters, it is still easy to find a person with a high tolerance to the high level of bitterness in certain beers, whereas some cannot even take a sip of those beers due to the incredibly bitter sensation on their tongue. So, why can bitterness expose such differences in taste sensitivities among us?

A wide variety of chemical compounds can be perceived by our taste buds as sweet, umami or bitter. It is easier to understand the sweet and umami story because our tongue needs to develop preferences towards a wide range of sugar and protein sources in order to recognize those macronutrients for basic survival. Similarly, we need to be able to sense a great variety of compounds as unpleasantly bitter, as those compounds are potentially toxic. Consequently, the bitter taste receptor genes diversified throughout evolution for the ability to recognize a diverse array of poisonous compounds that we may encounter in nature. In fact, at least 25 functional receptor genes are confirmed for bitterness, which suggests that the number of chemical compounds that can cause the bitter sensation is the largest when compared to the number of chemicals for other tastes.

The diversification of bitter taste receptor genes came from genetic mutations. These mutations either persist and are passed down through the generations or are killed off. For example, a mutation can shut off the genes that are responsible for detecting the potentially toxic PTC or PROP compound (previously discussed). If people are not exposed to these compounds in their living environment, for example due to subsisting on domestic agricultural products that do not contain such toxins, then the mutation in the gene will not be punished by the environment.

This means that the inactive status of the genes is passed on to the next generation, which leads to the insensitivity to PTC or PROP in future generations. By contrast, in a punishing environment where PTC or PROP-like compounds are ubiquitous in seemingly edible plants, such gene mutations would be killed off as without the ability to discern these dangerous toxins people with the mutation are less likely to be able to live on and procreate.

Because of the variations in bitter taste-receptor genes, taste preferences and dietary behaviour towards food and beverages vary among people. As discussed earlier, both tannins and alcohol are potential bitter compounds. Yet some people may not perceive the tannins and the alcohol as bitter at all. As a result, numerous studies on health behaviour found direct correlations between unresponsive bitter taste receptor genes and addictions to alcohol. Although not conclusive, it was frequently observed that a person with lower sensitivity to bitterness is also less sensitive to the 'burning' taste of alcohol. To those people, it is effortless to glug down drinks that contain high percentages of alcohol; such behaviour is even encouraged due to drinking cultures in many countries perceiving this ability as impressive and an indication of toughness. As a result, these people are more susceptible to alcohol dependence.

Other types of taste can also be largely influenced by genetics. For example, the composition and the production rate of saliva vary significantly among people. As explored in Chapter 5, a solution of salt has the sodium ion as the key stimulus of the salty taste. What's interesting is that sodium ions exist in our saliva too, but we do not perceive our saliva as salty due to taste adaptation (as discussed in Chapter 5). When we taste salty food and drinks, the concentration of sodium ions is usually higher than that in our saliva, so we can detect saltiness when eating and drinking. But wine generally has a low concentration of salt. Whenever two people debate on whether or not a wine tastes salty, it is likely that one of them cannot perceive the saltiness in the wine due to an equal or higher concentration of sodium ions in their saliva. And if the same wine has an obvious salty taste to the other person, it means that the person's saliva has a much lower concentration of sodium ions than the wine has, so their salty taste receptors are awakened by the higher level of salt in the wine. In the next chapter, we are also going to consider how the flow rate of saliva affects the astringent mouthfeel.

Gender and age

As we saw in Chapter 2, colour-blindness occurs more in the male population due to genetics. A lot of research has explored how gender and age influence taste sensitivities. With a few exceptions, the majority of research shows that female subjects tend to be more sensitive toward bitter, sour or salty tastes than their male counterparts. In terms of age, younger people have more sensitive taste perceptions than older people.

In the wine world, we see a clear difference in taste sensitivities. A famous example is a red Bordeaux fine wine – the 2003 vintage of Château Pavie. The same wine was reviewed by two of the world's most famous wine critics. One glorified the wine as brilliant, stating that it was among the greatest examples of the vintage, while the other found it completely unappetizing, with sweet Port characteristics. Other than personal experience and psychological factors, the tasting notes from the two critics over the years indicate that the one who disliked the 2003 Château Pavie is generally more sensitive to all tastes, including those sensations caused by wine components such as tannins and alcohol. The taste components are perceived as significantly more intense by this sensitive taster, to the point of being unappealing.

The wine example shown above is a luxury fine wine that is mainly sold based on its non-sensory qualities such as brand, reputation and classification. Therefore, the different taste preferences of the two wine critics were more commercially relevant (less relevant to the intrinsic sensory attributes when it comes to the marketplace). The media love such conflict of authoritative opinions and this leads to more brand exposure.

In contrast, for any wines that are designed to appeal to a large consumer base, the expertise in tasting fine wines matters less. Consumer sensory studies play a more important role in determining how to influence people's taste preference and gain large-volume and continuing purchases. We tend to find that many commercially successful wines sold in large volumes are sweeter, or more savoury, instead of being very acidic, astringent or bitter. These wines are designed to have the widest appeal possible, avoiding tastes that tend to be divisive among the general population, in order to maximize sales volume.

Yet, even if wines like white Zinfandel are sweet, with low alcohol and taste almost like fruit juice, the small population that can be qualified as 'supertasters' or hypertasters will still find the presence of alcohol taste sharp and unpleasant. Meanwhile, the hypotasters may only perceive such wine as somewhat sweet and would rather drink vodka with a high level of alcohol so that they can indeed feel a mild stimulating effect on the palate.

Final thoughts

A wine-tasting experience should not be painful, but there are compounds that trigger the bitter taste receptors, to which some people are particularly sensitive. This chapter gave an overview of the components that are the suspects that may contribute to bitterness. We should expect more concrete findings on bitter compounds in wine as research progresses. It is also just a matter of time before research unveils the role of fat as a potential sixth taste.

At the same time, advances in genetic research shed more light upon just how diverse our sense of taste is. It is easy to be wrapped up in our own sensory realm during wine tasting, but when sharing a wine with others it is important to acknowledge that they might be having different sensory experiences. To you, a glass of Barolo can taste as smooth as silk on the palate, while the person next to you may find the same wine painful to swallow.

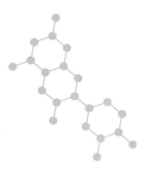

7
Tactile sensations and acquired taste

Beyond basic tastes

Other than the five basic tastes, there are multiple types of tactile sensations that happen when food meets our palate. Often, we use mouthfeel-related terms to describe those tactile sensations. These mouthfeel elements are perceived by the trigeminal nerve, not via taste buds. Such tactile senses function in the same way that other parts of our skin perceive touch–position and pain–temperature. In wine, the tactile sensations can be as important as tastes and aromas. The following sections aim to cover the majority of distinct tactile sensations – astringency, viscosity, irritation and temperature – during wine tasting.

Astringency

This section begins with a final major lesson on phenolics. It also explains what astringency actually entails and outlines what the other components of the section are as well as introducing tannins.

One of the major differences between a glass of white wine and another, of red wine, is that the latter generally contains tannins while the former has few. Aside from the anthocyanins, which are phenolics that impart the red colour, tannins are probably the most discussed phenolics in the wine world. Tannins exist mainly in grape skins and grape seeds. With a sufficient level of extraction, such as by fermenting the grape juice and the skins together in red winemaking, the resulting wine will contain the tannins that contribute to the astringent, puckering, or drying mouthfeel.

Phenolics lesson 3: tannins

The chemical structures of tannins were addressed in Chapter 1 in relation to the colour compound anthocyanins. Both tannins and anthocyanins belong to a group of phenolics which are flavan-3-ols (a sub-category of flavonoids). In grapes, the basic units of tannins are specific flavan-3-ol compounds such as catechin, epicatechin and epigallocatechin. Imagine that single-unit compounds such as catechin and epicatechin are like Lego building bricks, but with magnetic properties. These pieces tend to combine to form long chains. In technical terms, the individual flavan-3-ol units polymerize to form larger polymers, resulting in tannins. The process of polymerization can keep going as wine ages, resulting in large-size tannins or wine pigments (anthocyanin and tannin polymers as addressed in Chapter 1) which eventually become visible sediments in wine.

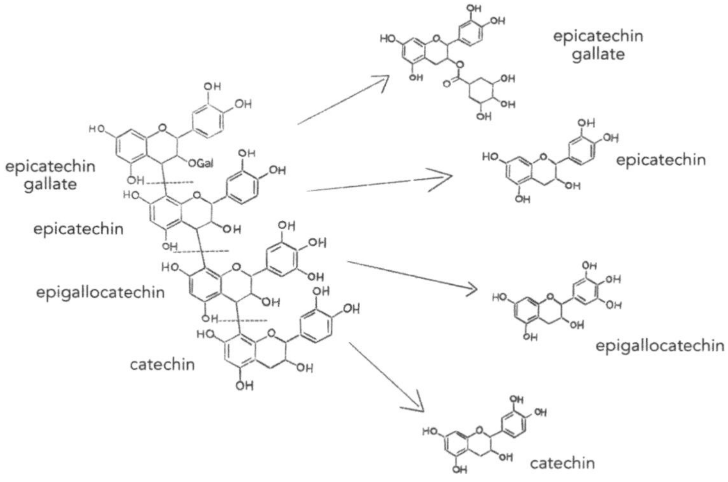

Figure 12: The polymerization of tannins. The components of tannins are single units of phenolics (flavan-3-ols). The same or different types of flavan-3-ols polymerize or condense together to form medium- to long-chain structures called tannins.

We will explore the complicated sensory impact of tannins later. But just by considering the chemical terroir, a veritable galaxy of possibilities becomes apparent. Constructing a tannin polymer of 10 units involves utilizing two building blocks, catechin and epicatechin. Through the principles of combination and permutation, the resulting polymer could (theoretically) have 524,288 distinct combinations. Indeed, one of the reasons for the infinite sensory experience of wine is the myriad variations of tannin polymers.

Tannins from oak can also be released into wine. These are called the ellagitannins and are ubiquitous in the plant kingdom. In the majority of winemaking grapes though, ellagitannins don't exist. The sensory properties of ellagitannins are not quite clear. They seem to give a similar astringent sensation, but some people claim that the exact mouthfeel is not the same as that given by grape skins and seeds. In white wines and in spirits like whisky, the oak tannins can add an obvious layer of astringent sensation. In red wines, studies have shown that oak tannins seem to have little impact on the mouthfeel. It seems that the highly concentrated tannins from grape skins and seeds mask the mouthfeel given by the much lower level of oak tannins.

The mysteries of sensory impact

The word 'tannin' was derived from the term tanning in leather production. In prehistoric times, our ancestors discovered that by treating animal skins with vegetable tannins (mostly from oak bark or from other woody materials), the skins or hides could be made more durable and better qualified to be worn as leather. The science behind this phenomenon is that tannins permanently decrease the active nature of proteins in animal skins, making resulting leather more resistant to decomposition. When tannins come into contact with our mouths, they reduce or remove the lubricating functionality of saliva proteins, resulting in the drying and astringent sensation.

Whether or not a person appreciates the tactile sensation of tannins depends on genetics (discussed in Chapter 6) and experience (to be discussed later). Personal preference aside, the astringency caused by tannins is an important attribute in drinks like wine, tea and coffee, so the producers of those beverages take great effort in understanding and managing the final sensory outcome of tannins. Today, some quantifiable measurements can help winemakers or consumers interpret the bigger picture of tannins in each wine. The total amount of tannins, along with other phenolic compounds, can be accurately measured. In general, the higher the concentration of tannins, the more astringent a wine can feel.

The composition of tannins in wine can be vastly different depending on grape variety and indeed there are a variety of tactile sensations across tannin-containing wines. However, even if the entire chemical make-up of a glass of wine is analysed, it is difficult to correlate the chemical data to what we should perceive on the palate.

Based on research in recent decades, it seems that certain parameters of tannins are associated with the tactile sensation on the palate. For example, some research has reached a consensus that the degree of polymerization is a reliable method when it comes to forecasting the mouthfeel of tannins for specific types of wines like Bordeaux reds (though a lot more research and data are required for validation). Each year, winemakers of certain premium Bordeaux wineries look at the average number of single units in a tannin polymer and correlate such measurement with historical data. For example, among those great vintages when the bottled red wines were highly praised by producers, professionals and consumers, the degrees of tannin polymerization are seen to be similar. Based on this observation and experience, those wineries can treat specific degrees of polymerization as magic numbers to guide grape growing and winemaking.

Not all research has reached the same conclusions. Some studies have found little correlation between the tannin mouthfeel and the degree of polymerization. The seemingly conflicting research conclusions can also be seen regarding the sizes of tannins and their sensory effect. Some research has shown that the larger a tannin polymer is, the more astringent a wine can feel. Yet other studies have observed the complete opposite.

One of the most important factors that lead to the different conclusions is the type of wine used in research. It was often observed in wine-related research that the data and findings vary depending on the wine samples used. This is called the matrix effect, which means that a target compound in wine is expressed in different ways due to the other compounds that exist. For example, two wines can have the exact same level of alcohol, but the tactile perceptions of alcohol can be completely different (one tastes smooth, while another has a burning sensation) because of the many other different chemical compounds in the two wines. Regarding tannins, the other components in the matrix such as alcohol, acidity and polysaccharides (to be discussed later in the section on viscosity) can all impact the overall sensation of astringency. Therefore, it is not surprising that tannins of the same degree of polymerization and/or of similar sizes have different tactile perceptions in different kinds of wine matrices.

Wines made from different grape varieties, in different vintages and using different winemaking techniques have different tactile profiles when it comes to tannins. In order to appreciate how tannins and other phenolic compounds lead to complexity on the palate, it is important to experience as many wines as possible during your lifetime (what a terrible burden that's going to be! And great news for your dental hygienist too). By tasting more wine varieties, especially by comparative tasting, you will develop your own sensory localizations of varied tannin profiles. For example, in the same area of production, the tannins of a Pinot Noir have a different mouthfeel to the tannins of a Cabernet Sauvignon. The same vintage of a Grenache Noir and a Nebbiolo should taste vastly different regarding the astringency. Although they use the same grape variety, a Merlot from the Bordeaux region will produce a different tactile sensation than a Merlot from California's Central Valley in terms of its tannin profiles. For anybody just beginning their wine-tasting journey, tasting grape varieties and regions of origin with such contrasting tannin profiles can be informative.

Last but not least, there is huge genetic variability in terms of how our palate perceives tannins. Studies have shown that people with more saliva flow perceive tannins as less astringent. The reason is quite straightforward: a higher amount of saliva means more proteins available to react with tannins, as well as to sustain the lubricating effect on the palate. This explains why highly tannic wines made from grape varieties such as Cabernet Sauvignon, Nebbiolo and Tannat can be consumed effortlessly by those people who salivate a lot, while others cannot tolerate the mouth-drying effect after the first sip.

Viscosity

In the same way that our fingers can feel the difference between water and oil in terms of their viscosity, our tongue is able to tell which wines are more viscous than the others. There are other terms that are more commonly used to describe the viscosity in wine, such as body (fuller or lighter), richness (richer or thinner) and weight (heavier or lighter). This section will primarily use the term body as it is widely used in the wine industry.

Many types of chemical compounds in wine can make the body fuller or lighter. One of the mechanisms that creates body is that certain compounds increase the density of wine as a liquid solution. Sugar, for example, increases the density of water so that wines with high levels of sweetness will feel heavy on the palate. Wines like Rutherglen Muscat and PX Sherry can contain more than 400 grams per litre of residual sugar, which feels almost sticky.

But density alone cannot explain the body of a wine. The major alcohol in wine is ethanol and ethanol solution has a lower density than pure water. However, wines with higher alcohol content seem to be fuller bodied in most cases. This is probably because ethanol can create other types of tactile experiences such as the burning mouth sensation (see Irritation, below). Experiments have proven that within a certain range, the increments of ethanol levels correspond directly to the increasing body of a wine.

Anecdotal experiences also tell the wine industry that compounds like polysaccharides and glycerol can give a richer mouthfeel, whereas bubbles and acidity make a wine lighter in body. But again, conflicting results have been seen in different studies. Glycerol, primarily produced by yeasts during fermentation, has been examined in some studies and its contribution to a fuller body and even sweetness in white wines was demonstrated, while other research concluded that the body (or viscosity) of a wine could not be improved by the addition of glycerol. Other elements like tannins have been discussed among wine professionals and researchers, who debate whether they increase or decrease the body of a wine. Some people consider that the texture of tannins makes wines have a fuller mouthfeel, while others argue that the astringency of tannins leads to a thinner tactile sensation.

In summary, the body or the viscosity of a wine is a tactile sensation on the palate that is influenced by a great variety of elements in the liquid. While there are limitations in the current research methodologies on the exact sensorial outcome of a wine's body based on chemical and physiological data, we are aware of the complex interplay of chemicals that leads to the matrix effect in wine. A common pitfall is to focus on one particular element and consider it as the sole contributor to a specific sensation. The body of a wine is always an overall tactile sensation on the palate.

Irritation

In food, the most prominent examples that cause irritation on our tongue are chilli peppers. The chemical compound capsaicin in chillies triggers sensations that are commonly described as spicy, hot or burning. Hypertasters, as described in Chapter 6, cannot stand even a tiny bit of chilli influence in food, whereas some tolerant people are addicted to the taste.

Wine does not possess capsaicin, but it has other compounds that can cause irritation on the palate. Alcohol, mainly ethanol, can lead to a burning sensation, particularly in hypertasters but even in hypotasters at a high enough concentration. Due to different sensitivities among people, the evaluation of alcohol by our palate can be quite a challenge in terms of reaching a consensus.

Bubbles, as they burst in our mouths, cause irritation. But this sensation can be pleasant to many consumers as it adds a refreshing and fun element while drinking. The mouthfeel of bubbles is highly dependent on the pressure in the bottle and the other compounds in wine. Many studies on sparkling wines, beer, cider, and other bubble-containing drinks have noticed significant positive or negative correlations between the alcohol content and the other components in these beverages such as acidity, phenolic composition, sugar, polysaccharides and proteins. It is important to keep in mind that the texture of the bubbles is down to much more than simply the bubbles themselves. Again, it is the complex matrix of the wine that shapes the overall mouthfeel of the mousse we perceive on the palate.

Temperature

In recent years, more and more research has been conducted to evaluate the effect of serving temperature on the taste of wines. We learn from experience that different temperatures have a great influence on how a wine tastes but there is no chemical or physical formula that can predict the exact change.

However, certain sensory studies have shown interesting results. In one experiment, different serving temperatures resulted in varied sensations towards sweetness and acidity intensities in white wines. Another study noted how the astringency, sourness and bitterness in a red wine were affected by different serving temperatures. As mentioned in Chapter 5 (recall our discussion on pKa and the lingering sour perception), the lasting effect of the sour sensation in wine should theoretically change under different temperatures. In Part Three, we will explore another property of wine that is influenced by temperature: its aroma.

For grown-ups

It has been proven that new-born babies inevitably react to a bitter taste negatively. As we grow up, our sense of taste develops and we learn to appreciate safe food and beverages with some bitterness. Coffee is the most obvious example of this acquired taste. Many of us initially bear with the bitter taste, perhaps to obtain caffeine to stay awake in our busy lives, or attempt to counteract it with cream and sugar. Over time, however, we come to enjoy coffee's taste and to appreciate its complex aromas. Thanks to the advanced technologies that ensure food safety, we don't need to rely on our taste buds to avoid poisonous consumption. As noted above, from an evolutionary point of view, it is not surprising that a portion of the population has a decreased sensitivity to specific taste compounds, as detecting them in our food is no longer a matter of life and death.

While genetics play an important role in sensitivities towards tastes, our minds can also significantly influence how we perceive tastes. We can overcome repulsive tastes when consuming food and beverages if we have reason to do so. For example, if we know a bitter medicine is good for us, we will drink it despite the bad taste it leaves in the mouth. We might even grow to enjoy some flavours we initially dislike (such as blue cheese), once we become comfortable with the fact that a given food should taste and smell this way and is safe to consume. Conversely, even if something tastes pleasant our minds can be repelled. For example, we are now aware of the consequences of too high a sugar intake and therefore if something tastes incredibly sweet, we are more likely to perceive that food as harmful despite its lure for our taste buds.

There is evidence of how diet affects our preferences in taste, especially during the early stages of life. For example, the increased acceptance of and preference for more heavily salted foods are positively correlated with exposure to salty foods during infancy. The learned response was also observed in the case of sweet taste. Children who become used to a highly sweetened diet tend to be vulnerable to overconsumption of sugar later on.

This is why there can be confusion or debate over the quality of food and beverages. For example, if the quality of pungent Roquefort cheese needs to be evaluated, the judges must accept and appreciate such cuisines, which may be perceived as unpleasantly smelly or pungent by novices. With increasing globalization, there are more and more cuisines available from around the world, with a range of tastes for our tongues to adapt to.

Wine is no exception to this, as most wines emphasize acidity, tannins, burn or other sensations that call for an acquired taste. During professional tasting events such as wine competitions, at least one judge in the group must be familiar with the style or the region of the wines in the line-up. When rating wines with scores, a wine critic should only evaluate the wines that he or she is very experienced with. For example, the retired wine critic Robert Parker (mentioned at the end of Chapter 3) specialized in Bordeaux wines but not Burgundy, therefore Burgundy wines were evaluated by the Burgundy experts in his team.

This is also why this book constantly reminds readers of the variabilities of sensory perception and the causes behind those variabilities. In most wine tasting scenarios, both our past and present can influence what we taste in the glass and whether we like or dislike it at a given moment. Before moving on to Part Three, which addresses the even more variable sense of smell, it is important to emphasize again that the true science behind wine tasting is far more complex than 'chemical A leads to sensorial response B'. Instead, matrix effects can strongly influence the sensory outcome and the situation of wine tasting is dynamic.

More importantly, this is why we should keep an open mind towards wine tasting. At different times in our lives, we may have entirely different perceptions of the exact same wine due to the possible changes in our physiological responses to tastes as well as the accumulation of experiences, whether or not they are directly related to wine tasting.

Final thoughts

I hope that these discussions touched every corner of your palate. Writing this chapter reminded me how little we know about how our palates experience tactile sensation but, at the same time, increased my appreciation of the complexity behind the mouthfeel of wines. Astringency, viscosity, irritation, temperature – these sensations 'scratch' our palate during almost every wine tasting. Sometimes, they might be perceived as unpleasant, but for the most part, they add a lot more fun and variety to the drinking experience.

Of course, there is a reason why wine is a beverage for adults, beyond health-based age restrictions. Complex and somewhat challenging tastes like astringency, richness or heat require knowledge and experience, which also builds upon our individual biological sensitivity. To most people, wine is an acquired taste. Or to put it another way: the more experience you have of food and drink in general, the more wines you are likely to appreciate. In the next part of the book we will discover that it is more than just the taste that makes you like or dislike a glass of wine: the smell on the nose probably matters just as much, if not more.

PART THREE
Smell

'Odours have a power of persuasion stronger than that of words, appearances, emotions or will. The persuasive power of an odour cannot be fended off, it enters into us like breath into our lungs, it fills us up, imbues us totally. There is no remedy for it.'

Patrick Süskind, *Perfume: The Story of a Murderer*

The sense of smell, also called olfaction, is much more significant than we might imagine. While it may be true that to survive in the modern world, the sense of smell is not as crucial as other senses, those who lose their sense of smell cannot emphasize enough how much of a negative impact the inability to smell has on their lives. Our olfactory bulb, which is the receptor for odour signals, is encircled by our emotional centre in the brain. Studies have demonstrated that significant smell loss is often associated with symptoms of depression among patients. According to some patients with smell loss, the whole world seems to become so much less 'colourful' without being able to perceive aromas.

It was rumoured that, when he was the most influential wine critic in the world, Robert Parker insured his nose for a million dollars. There are two implications behind this story: firstly, professionals in the wine industry are well aware of the importance of smell. As revealed later in this book, it is the nose rather than the palate that detects the majority of flavours when we eat and drink. Secondly, the physiological structure of olfaction is fragile. Quite a few viral diseases and a low level of concussion can lead to the temporary loss of smell, for variable lengths of time. In comparison, our palate is much more robust. Even if the tongue gets burnt or bruised badly, our taste can recover quickly as the cells of the taste buds replace themselves every one to two weeks.

Our sense of smell is far from well-studied because this subject is so incredibly complicated. Theoretically, we can differentiate a trillion different types of smells, even though our vocabulary for naming precise smells is very limited. Furthermore, we see even more subjective variabilities in the sense of smell (for example, through genetics) as compared to the other senses. Examples of variations in aroma perceptions are everywhere, such as how one person might love the smell of coriander (cilantro) while another detests it. Whenever there is debate during wine tasting, the different perceptions and interpretations of the aromas are likely to be the main cause.

In Part Three of this book I will present and explain to you the current scientific understanding regarding the smell of wines. This part examines the chemical origin of the different groups or types of aromas commonly described in wine tasting. Each chapter addresses a specific sensorial feature regarding our sense of smell.

Chapter 8 begins our aromatic journey with the attractive floral and fruity aromas usually associated with the plant world but also found in wines. These aromas are often brought out by compounds like terpenoids, esters and sulphides, which we look at in detail in this chapter. There is also an emphasis on volatility, which is a crucial chemical feature that allows a chemical compound to be detected by our noses. The chapter ends by emphasizing the importance of understanding the two modalities of smelling: orthonasal olfaction and retronasal olfaction.

Chapter 9 focuses on the vegetal and herbal aromas that can appear in wines. We revisit terpenoids and sulphides, as these two categories include compounds that are responsible for the green aromas of vegetables and herbs. There are other chemical groups that contain compounds with vegetal and herbal aromas, such as higher alcohols and pyrazines. Some of these compounds can contribute to a strong smell even at low levels. Therefore, the sensory topic of this chapter is about the odour threshold, a potentially misleading concept that needs careful interpretation.

Chapter 10 examines the gamey and musty aromas that are generally considered to be signs of faultiness in wine, especially if there is a high intensity of such odours. The aroma compounds associated with those smells can originate from the grapes, such as the methyl anthranilate compound responsible for the foxy smell of specific grape varieties. Various types of microbial contamination during wine production are the major source of faulty aromas. Hence we'll explore microbial taints in wine. Lastly, in this chapter we find out about the sensorial mechanism called odour adaptation, which makes foul smells seem to get weaker if we are exposed to them for a longer period of time.

Chapter 11 highlights the spicy and woody aromas that many plants possess. We don't usually eat spices and the woody parts of a plant on their own, but they play an important role in flavouring food and beverages. Although certain grape varieties do have the ability to produce spicy aromas, the use of oak in winemaking makes the most significant contribution to spice and wood-related smells in wine. We will look at these oak aromas in detail. This chapter emphasizes once again the significance of our genetics when it comes to the sensory side of things. Certain aroma compounds can be perceived by some but not by others. We will look into the implications of this 'specific anosmia' when it comes to tasting wine.

Chapter 12 concludes the world of wine aromas by considering whether the related compounds are chemically oxidative, reductive or neutral. Many compounds covered in earlier chapters of this book will be reviewed from the oxidative–reductive perspective. This chapter will also illustrate some other important wine aroma compounds that have not yet been addressed, many of which are relevant to how the oxidative or reductive approaches to winemaking result in the corresponding aroma features. We end our sensory journey through these myriad aromas by looking at just how powerful our sense of smell is.

8
A bouquet of flowers and a fruit bowl

Smell – a new language

We are surrounded by solid references for our sight: the sky is blue, the leaves are green, the wine is red. We are constantly exposed to concrete references for our palate: the honey is sweet, the lemon is sour, the wine is bitter. But when it comes to the sense of smell, we often become clueless in terms of recalling the best references for aromas. Most wine drinkers have not been taught to smell an apple and a glass of wine together and then decide if the wine has apple-like aromas. Even many well-trained wine professionals find it almost impossible to provide a good reference for the 'mineral-like' smell in wine, because the real minerals on our planet do not contribute to any smell.

Therefore, learning how to describe and communicate what you smell is like learning a new language. There is no established learning system for such a language and there are no rules in terms of how you want to express what you smell either. Hence, you may encounter lots of aroma descriptions out there which do not make sense to you at all (e.g. if one person says a wine 'smells like grandma's wardrobe'). The biblical Tower of Babel story seems a suitable analogy when it comes to the language of smell. Yet, we do not have to speak the exact same language to understand each other. Two tasters can either like or dislike the smell of the same wine. The happy face of one and the frowning face of another say it all. There seem to be underlying principles behind the mysterious sense of smell and they are understood better by scientific discoveries.

This chapter begins by introducing volatility as one of the most important chemical properties of aroma compounds. Then a large number of compounds associated with fruity and floral aromas are introduced. Lastly, the important mechanisms of orthonasal and retronasal olfaction are illustrated. Once you understand these scientific principles, you can get creative with establishing your own vocabulary for the world of smells.

Smells are 'volatile'

Lights hit our eyes to give us vision. Compounds in solutions contact our tongue to trigger the sense of taste. Those compounds that 'escape' into the air from solutions or solids travel to the receptors in our nose, triggering our sense of smell.

Most sugar compounds, as discussed in Chapter 4, do not contribute to any smell. This is because the molecules are tightly locked in water solutions without being able to escape into the air. This chemical property is described as non-volatile. By sniffing a glass of pure sugar-water solution, you will not be able to smell anything sweet, nor can you detect any other odour. Ethanol, on the other hand, is much more interesting as it is very volatile. Therefore, solutions with high alcohol content not only give burning sensations to the palate but also trigger our noses into action and have an odour. Volatility itself can be influenced by factors like temperature. The higher the temperature, the more volatile the compounds become. Hence, the increased alcoholic sensation on the nose when we drink warmed Japanese sake, compared to drinking it chilled or at room temperature.

In order to attract animals, plants make specific aroma compounds and concentrate them in the fruit. Throughout evolution, many types of fruit, such as wine grapes, developed thousands of different compounds that contribute to the floral and fruity smell. The result is the countless volatile compounds found in nature, thousands of which are found in wine.*

But you should not be intimidated by the sheer number of wine aroma compounds. This book is all about the scientific principles and the approach to understanding what we smell in wine tasting. There is no need to memorize each chemical compound without knowing its role in the chemical and sensorial properties in wine. In fact, plants are smart enough to figure out that to construct a diversity of odorants, the basic chemical units do not need to be complex. Therefore, in all the chapters on smell, it is my intention to equip you with a good understanding of the basic units of aroma compounds – you'll find that there isn't much that you need to remember.

Let's begin by exploring the three broad chemical groups of volatile compounds: terpenoids, esters and sulphides, which are responsible for the majority of floral and fruity aromas in grapes and wines.

Terpenoids: countless possibilities

Terpenoids are the largest and most diverse class of compounds derived from natural sources. Note that terpenes, a term more commonly used in the wine industry, are those terpenoids composed of simple carbon structures. In other words, terpenoids (more often used in this book) are more complex chemical forms of terpenes. Although they appear in numerous forms, the building blocks of terpenoids are simple five-carbon units. To plants, the terpenoids are the easiest constructions to complete as photosynthesis and metabolism are all centred around carbons (often just referred to as 'C' in the chemistry world). Yet, these carbon-based compounds are truly fascinating. Using carbon building blocks, plants are able to craft a near infinite combination

---

* A great resource for those who want to know more about the world of aroma compounds is the book *Nose Dive: A Field Guide to the World's Smells*, by Harold McGee.

of molecules, allowing each type of fruit to express distinct scents. Even before being converted to wine, different grape varieties have completely different constituent terpenoids. And even the same grape variety will create different combinations of terpenoids due to environmental variations (otherwise known as the natural terroir). As noted in the Introduction to this book, people often talk about wine being the most terroir-sensitive product, meaning wine tends to show the nuances of different places and vintages. This is certainly true when our noses detect the different matrices of terpenoids between two wines.

What's even more complicated is that the structure of terpenoids can be modified, even when just left in the bottle. During the long period of bottle ageing, the acidic environment of wine catalyses the structural change of certain terpenoids. For example, there is evidence that the floral-scented terpenoid linalool can be converted to the eucalyptus-like 1,8-cineole via chemical reactions in the pH environment of wines. This topic is still not well-studied, but we hope to address the subject further in future editions of this book.

Why do many terpenoid compounds smell floral and fruity? Again, this is because the fruit wants to send signals on the status of its maturity. Carbohydrates, like sugars, are not volatile and cannot be used as signals to gain an animal's attention from a distance. Therefore, the grapes either change colour to golden, pink or purple to get our visual attention, or they release aroma compounds like terpenoids to get noticed. Nowadays, some grape growers rely on not only the sugar level but also the concentration of specific terpenoids (such as beta-damascenone) to decide whether their berries are ripe enough for picking. Those growers know that the development of terpenoids reflects the grape's intention of telling us that it is ripe and delicious. But keep in mind that the smell of terpenoids is more than just fruit and flowers. As we will see in later chapters, certain terpenoids have spicy aromas and some even possess chemical solvent-like smells.

In this chapter, we will consider some of the most well-studied fruity and/or floral-smelling terpenoids in grapes and in wine.

There are two important aspects to keep in mind while reading about each terpenoid example:

1. Although each glass of wine contains multiple terpenoids, certain grape varieties are famous for having one or more terpenoid compounds that contribute to the most prominent types of aromas. Therefore, at this stage, research can only highlight the key chemical contributors to the major sensorial outcome via the sense of smell.

2. In many cases, one terpenoid compound corresponds to multiple types of fruit, floral and/or vegetal aromas. This is because, in nature, multiple flowers or fruits share the same terpenoids. In other words, most terpenoids are not unique to one given plant. The examples below will illustrate this phenomenon.

Linalool: citrus, lavender-like

Linalool is probably the king of terpenoids when it comes to its use across the world. It is a compound found in many products used throughout our daily lives, including cosmetics, perfumes, soaps and household cleaners. Why is linalool so widely applied to these products? It is simply because linalool smells good to humans. In nature, insects are also drawn by linalool and other terpenoids to help with the pollination of plants. Other than the pleasant smell, linalool has pharmacological properties, including antimicrobial, antioxidant, anti-inflammatory and anticancer. The more researchers have studied this compound, the more attributes they have revealed.

The smell of linalool reminds us primarily of citrus and lavender, as well as mint, laurels and rosewood. Of course, this is because all these plants are naturally abundant in linalool. Because of the ubiquity of linalool, it is difficult to find a single item in nature that pinpoints the aroma of this compound. Yet, we are familiar with the smell of linalool as it is present in lots of natural and synthetic things all around us. When you detect aromas that are citrus, or floral-like, think of linalool.

In the wine world, linalool-rich wines are often those made from the grapes in the Muscat family. If you've tasted a Muscat-based wine, it should be easy for your brain to connect the aromas to descriptors like orange blossom, acacia, elderflowers, lemon and orange. Smelling wines made from Muscat varieties is the best way of appreciating the fragrance of linalool.

Geraniol: geranium, citrus, rose-like

The name geraniol derives from the geranium, a genus of flowering plant that contains more than 400 species. Typical aromas of the species remind people of roses with a hint of lemon. This can be similar to the smell of linalool, but more rose-like, and it tends to be more pronounced in its aromas. Geraniol is as significant as linalool in our lives. Lots of essential oils contain a mixture of linalool and geraniol to create the most charming aromas. Not surprisingly, wines made from the Muscat family of grape varieties are equally abundant in geraniol.

Nerol: green, citrus, rose-like

The aroma compound of nerol must be included to complete the 'trio' of citrus, flower-like terpenoids. If you were to look at the chemical structures of linalool, geraniol and nerol, you would see they are astoundingly similar. In fact, these three compounds are isomers, which means that they have the same number and type of atoms, but the atoms are arranged differently. Therefore, this trio of terpenoids shares a lot of similar chemical properties and sensorial outcomes, yet we can detect slight differences among them on the nose. For example, nerol tends to be a bit more refreshing in its aroma, linalool contains an extra layer of spicy notes as found in lavender, while geraniol's aroma is the most rose-like.

Alpha-terpineol: pine, lilac-like

There is another isomer of the trio called alpha-terpineol. However, the chemical structure of alpha-terpineol is notably different when compared to the other three compounds. Perhaps this explains why alpha-terpineol has a more distinct aroma profile, reminiscent of pine and lilac flowers. This compound is often found in grapes with obvious floral characters, which once again includes the Muscat family of varieties, as well as Viognier, Riesling and Gewürztraminer.

Rose oxide: rose, lychee-like

Rose oxide, especially in the specific form of *cis*-rose oxide, is an aroma compound that gives memorable smells. In nature, rose and lychee are both rich in *cis*-rose oxide. Thus, ripe lychee fruit expresses a rose-like fragrance, while roses can remind people of lychee if they have ever tasted such fruit.

The Gewürztraminer grape variety, which has a high concentration of *cis*-rose oxide, is the most instructive variety for teaching new drinkers about the rose and lychee aromas in wine. Should you be presented with roses, lychee fruits and a glass of Gewürztraminer together, the magical smell of *cis*-rose oxide would connect these three seemingly distant sets of items together, setting up a relationship between them in your mind. Perhaps one of most intuitive exercises in developing a vocabulary of wine aromas is to experience the rose and lychee smells in Gewürztraminer wines.

Beta-damascenone: fruity, floral, cooked apple-like

Many research papers have reported some concentration of beta-damascenone in wine. However, its direct impact on our senses is not clear. This is probably because beta-damascenone exists in so many fruits and flowers that we can only think of something generically fruity and floral when smelling this compound. Recent studies have found that beta-damascenone can enhance the overall fruity aromas in wine. Because of its important role in shaping the general fruity and floral sensations, wine and beer producers nowadays perform lots of trials on enhancing beta-damascenone in their final products. As discussed previously, certain grape growers use the concentration of beta-damascenone

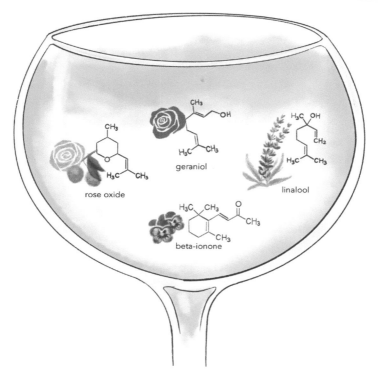

Figure 13: Examples of terpenoids

as a useful index to evaluate the ripeness of grapes prior to harvest.

Beta-ionone: violet, rose, raspberry-like

Many types of flowers, such as violet, as well as fruits like raspberry, have high concentrations of beta-ionone. If you browse tasting notes on wines made from the Pinot Noir variety, descriptors like violet, rose and raspberry often appear. Indeed, Pinot Noir wines generally have higher levels of beta-ionone compared to many other varieties. Another example is the Portuguese variety Touriga Nacional, which can have an even higher concentration of beta-ionone. Hence the smell of violet is quite obvious in Portuguese red wines that contain Touriga Nacional.

Terpenoids not only relate to floral and fruity aromas. In Chapter 9, we will learn about the terpenoids which give green, vegetal-like aromas, and in Chapter 11 we will see that the aromas from certain terpenoids are more exotic: aromas that remind us of pepper or petrol will be discussed. From all of this, it should be quite clear that terpenoids permeate every corner of nature. Other than plants, microorganisms, animals and humans also produce terpenoids that are important to them (or to us). It seems that terpenoids can be a powerful language in the form of aromas that every living being can use to understand and communicate with each other. We could discuss terpenoids endlessly, but we need to move on to the next important aroma group – the esters.

Esters: fruitier than fruit

The central element of metabolism is sugar, which consists of carbon plus hydrogen (H) and oxygen (O), hence the name for sugars: carbohydrate. Thus, it is not surprising that plants can make aroma compounds that contain a bit more than simple carbons. In fact, quite a few terpenoids (such as the trio of linalool, geraniol and nerol) also include the element oxygen. Now, let's look at another important class of generally fruit-like aroma compounds called esters, which are based on carbon, hydrogen and oxygen.

During the metabolism of plants, there are considerable amounts of intermediate and end compounds formed, such as different types of acids. Some of those acids can be modified to become esters in fruit. For example, isoamyl acetate, with a banana-like smell, is an ester found in many fruits. Although the banana fruit uses complex pathways to synthesize this compound, isoamyl acetate can be synthesized simply through reactions between isoamyl alcohol and acetic acid via chemical methods. Isoamyl acetate can also be reverted to isoamyl alcohol and acetic acid. Therefore, almost all esters can be deciphered by their names. Given the naming pattern just described, you can perhaps now tell, for example, that the chemical structure of ethyl butyrate is essentially formed by ethanol and butyric acid.

But plants only produce a small proportion of the esters found in wine. The majority of fruity aromas caused by esters are produced by the masters of ester engineering – the yeasts. During the fermentation process, yeasts consume sugar as their energy source and turn sugar into alcohol. But there is also a huge pool of alcohol and acids that yeasts like to use to create esters as side-products. Some scientists suspect that yeasts do this because it is a process of detoxification (converting the more toxic alcohols and acids into the less toxic esters). Whatever the true reason is, we could not thank yeasts enough for giving us not just one gift but two: alcohol and the lovely fruit-scented esters.

People in the wine community often describe the fruity aromas in wine by specifying what kind of fruit the wine smells like. For example, this Chardonnay smells like pineapple and that Pinot Noir has strawberry-like aromas. Because of this, flavour chemists try to match specific ester compounds found in both a specific wine and a certain type of fruit. Yet, no wine has a matrix of esters that are a match to those found in any specific fruits. Again, just like terpenoids, it is the exact *combination* of esters that do or do not resemble the aromas of a specific type of fruit. One ester compound can remind people of more than one type of fruit. The following are some aroma characteristics that are associated with certain esters which are shared in fruits and in wines (note the overlap of esters between two or more kinds of aromas):

- Apple-like: ethyl butanoate, 2-methylbutyl acetate.
- Pear-like: ethyl decadienoate, isoamyl acetate.
- Peach, apricot-like: hexyl acetate, ethyl hexanoate, gamma-decalactone.
- Banana-like: isoamyl acetate, ethyl butyrate.
- Strawberry-like: methyl butanoate, ethyl butanoate, methyl hexanoate, ethyl hexanoate.

Moreover, the smell of an ester can change at different

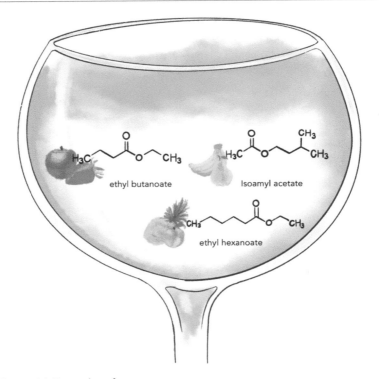

Figure 14: Examples of esters

concentrations. The most typical example is ethyl acetate, which possesses a generic fruity smell at lower concentrations. Yet, at high concentrations, it can smell like chemical solvents or nail polish remover, which is (unsurprisingly) considered a fault in wine.

What can make things even more complicated is that with the exact same concentration, a certain ester can give different aromas in different matrices of liquid or solids. If we apply the same formula of esters giving strawberry aromas in a gummy bear to a glass of wine, the wine may not smell like strawberry at all. Therefore, as the discussions of aromas go further, you will increasingly appreciate that it is almost impossible to precisely predict the fruit aromas in a wine based on chemical data; there are simply too many variables at play. The important takeaway is that we should never assume the aroma of a certain terpenoid or ester is the same in all kinds of food and beverages.

The concentration of esters in wine can increase or decrease over time. As discussed previously, the pure chemical reactions between acids and alcohols can form esters; this process is known as esterification. In certain conditions, esterification happens during a wine's ageing as there are many types of acid and alcohol present in wine. On the other hand, the acidic environment often supports the hydrolysis of esters, which means that the esters break down and revert back to the corresponding acids and alcohols. Although both esterification and hydrolysis happen during ageing, hydrolysis tends to be dominant in wine. That is why in most wines, the fruity aromas from esters have a tendency to become weaker as the wine ages.

Sulphides: love or hate

You may have noticed from the previous sections that certain terpenoids and esters do not always have pleasant smells. Such duality is much more obvious in the world of sulphides. The perfect example is the durian fruit, which is filled with sulphide compounds. Some people are addicted to the smell, while many people consider durian the most unpleasant smelling fruit.

The essential element for sulphide compounds is sulphur (S). In Chapter 12, we go into detail about sulphides and sulphites, which are easily confused with each other. This chapter makes a relatively simple introduction to sulphides by focusing on the fruit aromas that are caused by these compounds and whether those smells are enjoyable or awful. Note that a sub-group of sulphide compounds is called thiols, a term that is often used in the scientific world of wine.

Grapes generally do not produce many sulphides as aroma compounds. But certain grape varieties under specific conditions can synthesize precursors, which are chemical compounds that will be modified or converted to the target compounds we care for. In other words, certain sulphides in wine are formed by yeasts as they consume the corresponding precursors and turn them into the

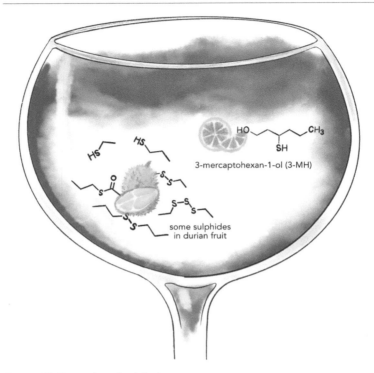

Figure 15: Examples of sulphides

sulphide aroma compounds. For example, the signature grapefruit-like aroma in New Zealand Sauvignon Blanc wines is caused by thiol compounds like 3-mercaptohexan-1-ol (often called 3-MH). In Sauvignon Blanc grapes, 3-MH appears in quite low concentrations, however, precursors such as certain unsaturated fatty acids and glutathiones can be more concentrated in this grape variety. Those precursors will eventually be converted to 3-MH after alcoholic fermentation.

A common sensory feature of sulphides is that we are extremely sensitive to them. This is because sulphides can be harmful to our health. For example, the simplest form of sulphide – hydrogen sulphide – is toxic at certain concentrations. Animals release hydrogen sulphides as gas to release these poisonous waste products from the body. Decomposing organisms harbour microbes that generate hydrogen sulphides and other sulphides, which is why hydrogen sulphide emits an odour reminiscent of rotten eggs. Those sulphides are synonymous with mortality. As a result, our nose has evolved to be on the alert for these chemicals and detect sulphides even at low concentrations.

On the other hand, some sulphide compounds can be harmless and even extremely beneficial to us. Once we figured out that coffee, which has a large pool of thiols, is safe to consume, many people came to greatly enjoy the smell of it. The reason is that the many forms of sulphides in coffee stimulate our sense of smell in various ways, resulting in an aroma complexity appreciated by humans. As another example, the seemingly dangerous hydrogen sulphide, when present at the right concentration level in our bodies, sends signals to regulate many biological activities, such as modifying neuronal transmission and protecting tissues from oxidative stress. This love–hate relationship between sulphides and humans will be discussed further in Chapter 12.

I hope that what we have discussed so far has helped you to better appreciate how plants, microorganisms, animals and humans interact with each other via volatile aroma compounds. We are unconsciously receiving important odour signals constantly, but only in recent decades have scientists started to pay more attention to the significance of our sense of smell. At least in the wine industry, people tend to have many more opportunities to focus on the wonderful smells in nature, as they need to correlate them to the similar aromas in wine. Those who are praised for having 'a good nose' in the wine industry were not necessarily born with it. They have simply gained more experience by actively smelling and tasting all kinds of substances around them rather than letting the aroma signals slip through their brains.

Two ways of smelling

Smell is probably the only sense that has two pathways. When we sniff an object, our nose senses that object from the outside, from the aromas it is giving off. When we put that item in our mouth, our sense of smell can also detect what is in our mouth, due to aromas making their way from the palate and into the nasal passages located behind the roof of the mouth. Although everyone will have experienced some loss of 'taste' when having a cold, most people are not aware of the fact that the inability to taste food during sickness is related to the blockage of pathways for smelling. When we consume food and drink, what we often perceive as 'tastes' are in fact aromas from the palate that reach our olfactory receptors through the nasal passages. This perception, called retronasal olfaction, can be pictured as an aroma compound travels from our palate and passes through the back of our mouth to the back of the nasal cavity, ending up on the nasal receptors that detect the aroma. This is why most food 'tastes' more than just sweet, umami, sour, salty or bitter on the palate: our nose also receives signals while eating and drinking. If a person loses their sense of smell from diseases such as COVID-19, most food becomes 'tasteless'.

The graphic below demonstrates the two pathways of smelling: orthonasal olfaction and retronasal olfaction. Orthonasal olfaction is more intuitive as we actively or passively sniff things even from birth. Retronasal olfaction was not well understood by researchers until recent decades, but easy experiments can be done to trace the smell from within the mouth. When eating chocolate, use a nose clip to close your nose, then you will find out that the chocolate

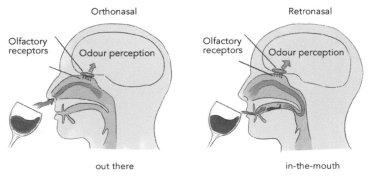

Figure 16: Orthonasal and retronasal olfaction

seems to lose all of its 'flavours' in the mouth. This is because most of the 'flavours' of chocolate are volatile aromas that are detected by our retronasal olfaction. Since wine has so many volatile aroma compounds, losing the ability to smell will make any wine taste nothing but sour, astringent or alcoholic.

To avoid confusion, throughout this book (as noted in Chapter 4), the term 'taste(s)' refers to the five basic tastes detected by our tongue and we use the terms 'aroma(s)', 'odour(s)' or 'scent(s)' for both smells received through our nose – orthonasal olfaction – and retronasal olfaction. 'Flavour', which will rarely be used (due to the confusion it might cause), is the interpretation made by the brain of the components 'aroma', 'taste' and 'mouthfeel', either individually or together.

Studies have demonstrated that our brain responds to orthonasal and retronasal olfaction differently. Moreover, different foods cause different brain activities. For example, chocolate gives more intense brain activity retronasally, while lavender triggers more of the orthonasal responses. It is hypothesized that because chocolate is considered a consumable food, our retronasal olfaction is more sensitive to produce the desire to eat chocolate. Lavender is less of a food but more of an aromatic object in nature so the orthonasal pathway is more active. In most cases, retronasal olfaction activates many more parts of our brain, despite some aroma compounds being more easily detected via the orthonasal pathway. Therefore, to have a comprehensive experience with aromas in food and drinks, we should combine smelling and tasting.

In wine, the difference between orthonasal and retronasal olfaction can be quite obvious. For example, the bouquet one might discover when smelling wine directly may differ from that experienced when tasting and 'inhaling' its scent retronasally. One major cause of such difference has to do with the volatility of the aroma compounds. For instance, not all terpenoids in grapes and in wines are volatile. In fact, most of them are attached to other compounds like sugar, keeping them non-volatile. This is because plants have worked out a mechanism to store the aromatic terpenoids and release them gradually into the air. The fruit doesn't want to give out all of its aroma at once and therefore immediately lose attraction to animals during the season of dissemination.

The most common combination of a non-volatile compound and a terpenoid is a glycoside, which consists of a sugar molecule attached to a terpenoid compound. The bond within the glycoside can be broken quickly by enzymes so that the terpenoids are released as volatile aromas. During fermentation, the enzymatic reaction within the yeast cells cleaves the bond, releasing quite a lot of terpenoids. This is one of the many factors that make the aromas of wine more complex than the aromas of grapes. During a wine's ageing process, terpenoids can decrease or increase in concentration, or transform into something else. If the majority of terpenoids in wine are released to become volatile during winemaking, those aromas tend to be lost quickly. This can explain why certain wines smell very aromatic in their youth but become bland and dull in a short period of time. However, if there are more terpenoids locked in the non-volatile, sugar-attached form, the aromas of those wines may not be very expressive at the beginning, but will tend to smell more complex after ageing. These wines may also be more aromatic retronasally than orthonasally. This is because our saliva contains enzymes that can break the bonds between sugars and terpenoids. This is why when we chew on the skins of grapes like Muscat, they become more aromatic in the mouth as the saliva breaks apart the sugar molecule and the terpenoid. If the concentration of enzymes in our saliva is high, we are likely to perceive far more aromatics on the palate than we will by sniffing a wine in a glass.

Final thoughts

Whether you deliberately use orthonasal olfaction to inhale the aromas from a wine glass, or unconsciously activate your retronasal olfaction by tasting the liquid in it, your nose will be filled with myriad volatile compounds when drinking wine. I hope this introduction to terpenoids, esters and sulphides gave a pleasant impression of the aromatic world. In the next chapter, we will continue to discover more terpenoids and sulphides, as well as encountering some other aroma compounds, which give us more than just fruity and floral scents.

9
Sniffing out the green aromas

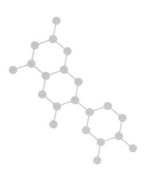

Green aromas?

Although chlorophylls in plants contribute to the green colour of plants, they are not volatile and therefore do not give off any smell. Yet we often use terms like vegetal, herbal, grassy or herbaceous to describe the aromas that remind us of the green parts of many plants. Although those 'green' odours do not come directly from chlorophyll, they are closely associated with it. Many interesting aroma compounds are predominantly produced within the green parts of a plant.

Some wines can express vegetal and herbal aromas. One of the goals of this chapter is to present the volatile compounds that are associated with the green aromas in wine, including our old friends terpenoids and sulphides, as well as some new introductions such as higher alcohols and pyrazines.

In the wine trade, it is not uncommon to find some people who are very sensitive to certain green odours in wine and others who are not. Various measurements of sensitivity to a compound are discussed in this chapter, especially regarding the careful use of the concept of odour threshold.

Terpenoids and sulphides continued

Since esters, terpenoids and sulphides exist ubiquitously in the plant kingdom, vegetables – much like fruits – are also masters at producing multi-functional compounds. Therefore, it is not surprising that certain wines express aromas that can remind us of specific vegetables, grasses or herbs.

Esters are mostly found in fruits and they are rarely concentrated in other parts of the plant. For this reason, the vegetal and herbal aromas in wine tend to come from compounds other than esters. Or, to put it more accurately, when we smell esters, the aromas are rarely, if ever, associated with the non-fruit parts of any plants. Esters tend to smell somewhat sweet and attractive to us because fruits produce esters to indicate their ripeness, as we discussed in the previous chapter.

Unlike esters, terpenoids and sulphides can be potent anywhere in a plant, and not all of them function as signals of ripeness. Instead, many terpenoids and sulphides serve as part of the defence system of a plant, in order to protect roots, stems, leaves and flowers, as well as fruits at earlier stages of growth, from damage by insects, animals and microbes. As a result of evolution, those aroma compounds are created to have an off-putting smell to repel other creatures. No wonder quite a few people dislike vegetables and fruits that contain certain terpenoids or sulphides.

For example, celery contains a few terpenoids that are not favoured by certain populations. A more obvious example is the Brussels sprout, which is rich in sulphides (such as hydrogen sulphide). Most people are not drawn to the aromas of Brussels sprouts, so chefs often put a great deal of effort into covering or balancing the smells of the sulphides during cooking. In contrast, certain green aromas can contribute to an aroma complexity that is desirable. For instance, the terpenoid compound citronellol exhibits lemongrass-like aromas which give wine and food a refreshing vegetal fragrance that is generally appreciated in the culinary world. Another example is related to one of the most essential components in beer – hops. Humulene (a terpenoid) is the key aroma compound in hops, which gives the distinctive green smells in beer.

In terms of sulphides, an example is the compound 4-mercapto-4-methylpentan-2-one (4-MMP in short) which produces

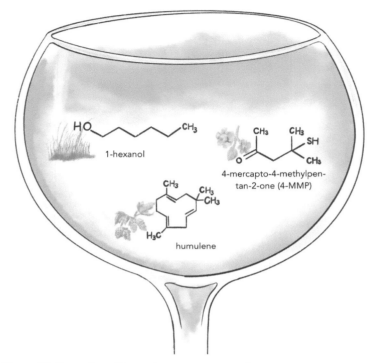

Figure 17: Examples of 'green' aroma compounds

aromas reminiscent of the boxwood shrub in wines made from Sauvignon Blanc grapes. If you don't know what boxwood smells like, it might help you to know that at certain times of the year, these evergreen shrubs release 4-MMP and other sulphides which remind some people of the smell of cat urine. Now you can understand why the term 'cat urine' appears from time to time in the tasting notes of Sauvignon Blanc wines.

Higher alcohols

The vegetal or herbal aromas can also come from compounds known as higher alcohols. The term higher alcohol does not imply that such compounds are superior to ethanol. In this case, 'higher' means 'more' in chemical terms. Chemically, ethanol has two carbons (methanol has just one), but higher (or long-chain) alcohols have more than two carbons.

In wine, the higher alcohol that is most relevant to grassy odours is 1-hexanol, which contains six carbons. This compound smells like grass, especially cut grass, because grass produces a significant amount of 1-hexanol. Like terpenoids and sulphides, the smell of 1-hexanol can be either loved or hated by humans depending on personal preference. In many cases, we can tolerate or even appreciate that grassy aroma from 1-hexanol. For example, the key aroma compound of olive oil is also 1-hexanol which gives an invigorating herbaceous scent which is appealing to most people.

Cut grass smells much more pungent than intact grass, not only due to the greater quantities of 1-hexanol produced when the grass is under the stress of being mowed, but also because some 1-hexanol molecules are oxidized to become *cis*-3-hexenal, which is an aldehyde compound (oxidation and aldehydes appear in Chapter 12) in which we smell more pronounced grassy aromas. While many people appreciate the smell of freshly cut grass, most of us are not drawn to such strong grassy smells in our food or drink and we are unlikely to develop an appetite for cut grass.

While many grape varieties can produce higher alcohols like 1-hexanol under environmental stress, Sauvignon Blanc is the most powerful producer of those compounds. As a result, the grass aroma appears very frequently in tasting notes for many Sauvignon Blanc wines from all over the world. While this aroma appears to be unappetizing to some consumers, it gives a vibrant and refreshing impression to many other wine lovers.

Pyrazine

The compounds 1-hexanol or *cis*-3-hexenal are not the only ones responsible for the herbaceous smell of Sauvignon Blanc wines. A category of aroma compounds called pyrazines is among the strongest contributors to the expression of green odours. Chemically, the pyrazine compounds introduce another chemical

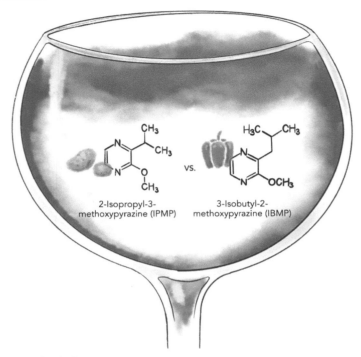

Figure 18: Methoxypyrazines

element that has not been covered yet. This element, nitrogen (N), is as essential as carbon, hydrogen and oxygen to any living organism, because nitrogen is the building block of the 'amino' part of amino acids, which are used to form proteins. With easy access to this abundant element, the plant wonders: why not use nitrogen to create some interesting aroma compounds?

To date, the most well-studied pyrazines in wine belong to a sub-category called methoxypyrazine. There are two major types of methoxypyrazine in wine: 3-isobutyl-2-methoxypyrazine (IBMP) and 2-isopropyl-3-methoxypyrazine (IPMP). The former, IBMP, smells like green bell pepper. The latter, IPMP, has a mixture of herbaceous, pea, earth and raw potato-like odours.

Grass, and the green parts of many plants, produce specific pyrazine compounds to defend themselves against herbivorous insects. Therefore, the smell of pyrazines tends to be 'unattractive' and can be very strong even at low concentrations. For example, we are particularly sensitive to IBMP because this compound signifies under-ripeness, inedibility or even toxicity to animals and humans. It is just like what we have discussed before regarding the aromas of certain terpenoids and sulphides; they usually signal that something is poisonous and thus we evolved to become sensitive to those smells to avoid danger. Research has shown that IBMP is detected by most people at the parts per trillion level, which means that if any chemist put one drop of pure IBMP into an Olympic-sized swimming pool, the entire pool would reek with the pungent smells of green bell pepper to most people.

'Bordeaux' varieties like Cabernet Sauvignon, Cabernet Franc, Carménère, Merlot and Sauvignon Blanc are capable of accumulating methoxypyrazines in the grape skins, while most other grape varieties have pyrazines in other parts of the vines but not in the berries. Wines made from these Bordeaux varieties can certainly exhibit green aromas, but they are in the matrix of many other aroma compounds together, so that the vegetal and herbaceous aromas are unique and complex in most wines. For example, some wines made from Carménère show more green bell pepper aromas (most likely dominated by IBMP) while some others smell distinctly canned pea-like (most likely dominated by IPMP). For Sauvignon Blanc wines, depending on the concentration ratio of IBMP and 1-hexanol/*cis*-3-hexenal, the green bell pepper and the cut grass aromas coexist to add to the complexity of vegetal, green sensations.

Aromas from external sources

Vegetal and herbal aromas can come from materials other than grapes (the wine industry uses the acronym MOG – **M**aterials **O**ther than **G**rapes). For example, if a vineyard is located close enough to a grove of eucalyptus trees, the MOGs will be the leaves of those eucalyptus trees, which release volatile terpenoids like eucalyptol that will settle onto the grape skins. Consequently, the wines will express minty, eucalyptus-like aromas due to the eucalyptol from the environment. Some wines show significantly more vegetal aromas due to the inclusion of stems on the grape clusters as MOGs. Winemakers may deliberately include stems during the fermentation process, especially for red wines made from varieties like Pinot Noir, Gamay and Syrah. This winemaking technique can add complex aromas that are often described as dried herbs or herbal spices.

Another example of MOG does not come from plants, but from certain insects. The Asian lady beetle (a type of ladybird now also widespread in North America and Europe) can be accidentally introduced into the fermentation tank if it gets into the grape batch. These lady beetles secrete methoxypyrazines as a defence mechanism (yes, insects can produce pyrazine compounds too) and can 'taint' the wines with aggressive green bell pepper and asparagus-like odours. You may have smelled this yourself if you have ever had a lady beetle or ladybird on your hand and it emitted this pungent substance.

Desirable or not?

There is debate in the wine industry regarding whether or not vegetal and herbal aromas are desirable. As you may have already worked out from what we learned about sensorial experiences in earlier chapters, the acceptance of green aromas is highly dependent on a person's genetics, culture and previous sensory experience. There are people who adore the green bell pepper aromas in wine. To them, those green notes add to the complexity of a wine. But not everybody likes them. For example, the US market, in comparison to other wine markets, has a reduced tolerance to any green bell pepper aromas in red wines. Just a hint of those vegetal odours may cause a wine to be rejected by North American consumers.

Volatility and sensitivity

As we transition from the chemical world to the sensory realm, there is one connection that's worth mentioning. Although we can be more sensitive to the smell of one compound over another based on our genetics, the chemical nature of aroma compounds can also play an important role in how easy it is for us to detect them on the nose. The rule is simple: if a compound has a higher level of volatility, it will travel to our olfactory receptors faster, increasing the ease with which we smell this compound.

Pyrazines, sulphides and other compounds like 2,4,6-trichloroanisole (see p. 151), are highly volatile compounds. If those compounds exist in a glass of wine, we can smell them, even from far away. On the other hand, the aroma compounds that do not escape easily from the wine in our glass are hard to detect. That is why we should remember to swirl the wine in the glass before smelling it – swirling helps these aromas become more volatile.

Odour threshold

The concept of sensory threshold is described as the lowest level of stimulus that can be perceived. In most cases, the wine industry cares about detection threshold, which is the lowest concentration that allows individuals with a normal sense of smell or taste to detect the presence of a substance. For example, if the threshold of an aroma compound is 2 milligrams per litre, below such concentration, the compound is not detectable. Therefore, if the threshold of an aroma compound is given a low value, it means such a compound can be easily detected by people at a low concentration. Another type of threshold which can be relevant to wine tasting is the difference threshold, which means the weakest concentration of stimulus that can make people perceive the same chemical compound with a different sensory outcome. A good example was discussed in Chapter 8, where we noted that the smell of the aroma compound ethyl acetate changes from 'fruity' to 'nail polish remover-like' at different concentration levels.

There is another commonly used index or calculation of sensory threshold called the Odour Active Value (OAV). The OAV is calculated by dividing the concentration of the aroma compound by the sensory threshold measured. This value may indicate the chemical compounds that contribute to the most potent aromas. In theory, the higher the OAV, the easier or stronger it is for a human nose to detect the compound. For example, in a study on the aromas of wines made from the Gewürztraminer variety against the Scheurebe variety, *cis*-rose oxide was concluded to be the major contributor to the much stronger lychee rose-like aromas present in the Gewürztraminer variety due to the significantly higher OAV of the wine sampled.

However, any reported threshold values can only be used as references and must be treated with great caution. We should not easily trust a specific number or range of threshold values for many reasons. Here are the two major considerations when reading about sensory threshold:

1. **Genetic difference means threshold values can never be absolute.** There is more on this topic in Chapter 11, but as we have seen in Chapter 6, people's sensitivities towards a specific sensory stimulus such as bitterness can vary a lot. For example, the threshold level of IBMP was described as very low on page 141, but there are a few people who may not be able to perceive the green bell pepper-like aroma of IBMP until its concentration increases tenfold.

2. **Threshold values depend on the matrix that the target compound was in.** This is the main cause of the different values of the threshold reported in different research papers. For instance, in one study the OAV of the ester compound isopentyl acetate was measured at 85.2 in a Merlot wine from Manasi County in Xinjiang Province, China, while the isopentyl acetate in a Merlot from Shacheng County in Hebei Province, China had an OAV of merely 7.3. The differences in threshold values can be much more dramatic in wines made from different grape varieties. Therefore, a threshold value must be checked against the beverage or food matrix that the compound is in.

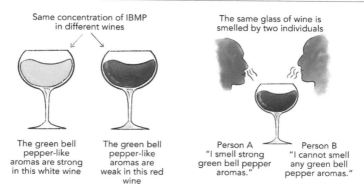

Same concentration of IBMP in different wines

The same glass of wine is smelled by two individuals

The green bell pepper-like aromas are strong in this white wine

The green bell pepper-like aromas are weak in this red wine

Person A "I smell strong green bell pepper aromas."

Person B "I cannot smell any green bell pepper aromas."

Figure 19: The unpredictability of thresholds

It is also worth addressing another complication in relation to threshold, which relates to the changes of detectable sensory properties over a period of time. Using sensory methods that evaluate the intensity of a specific sense or the temporal dominance of sensations over time, it is not uncommon to find that the perception of certain compounds can be either below or above threshold at different points during the consumption of food or beverages.

For the purpose of reference, there are two chemical compounds in the wine world that are famous for their extremely low sensory thresholds. One is 3-isobutyl-2-methoxypyrazine (IBMP), as discussed previously in this chapter, and the other is 2,4,6-trichloroanisole (TCA) which will be described in the next chapter. These two compounds have a detection threshold of 1 to 50 nanograms per litre. Here, the unit of nanograms per litre (which is also expressed as and equivalent to parts per trillion) is scarily small. Imagine looking at a photograph of all 8 billion people in the world and trying to make out the fingernail of one particular person in the entire crowd – that is how precisely our nose can perceive around 50 units of IBMP out of a trillion other compounds in one glass of wine. In such cases, the general values of the threshold can give us a good sense of the sensorial power of certain compounds even if the values can vary.

We are also sensitive to sulphides (though the general thresholds of sulphide compounds are lower than those of IBMP by a magnitude of 1,000), as mentioned earlier and discussed in detail in Chapter 12. This is because most sulphur-containing compounds are toxic to living things (see Chapter 8). In the wild, they are signs of rotten food and decaying corpses, which are going through the process of decomposition by the action of either internal enzymes or external microorganisms. Those volatile compounds released by the decomposition process are filled with sulphur- or nitrogen-containing compounds, suggesting death and therefore possible danger. We are sensitive to those odours for our own protection, so we keep a distance from both the rotten and potentially toxic bodies and whatever might have caused the death. On the other hand, scavengers like vultures who feed on dead organisms will find the decaying smell very attractive.

Lastly, it is worth mentioning that a person's sensory threshold may be changed through training. Research has observed that those who received tuition in wine tasting and winemaking became more sensitive to aroma compounds like diacetyl (which gives butter-like aromas) and mixed ethylphenols (some of which can give off-odours that will be covered in the next chapter) in certain wines. It is possible that exposure to those compounds can activate the genetic expression of specific odour detection.

Another reason could be the simple fact that we are naturally able to perceive certain odours, yet without training, our brain does not associate the perceived odours with specific items, experiences and languages. In this case, even if a person's odour receptor is activated by an aroma compound, their brain cannot give a response to it. Think of it like putting a textbook in front of a child who hasn't yet learned how to read. The child can perceive the words visually, but he or she will simply ignore them as the shapes on the page do not mean anything to them. Only after training in reading, will the words will become 'detectable' and meaningful.

Final thoughts

Terpenoids, sulphides, higher alcohols, pyrazines – our list of aroma compounds is expanding. But their origin is simple: nature intends our nose to receive signals, whether they are interpreted by our brain as 'green' or not, whether they are pleasant or irritating. This chapter also demonstrated that the same odour signal can be strong or weak, depending on the type of wine that carries that signal (the overall chemical matrix can greatly influence the signal) and the individual who detects it (given that sensitivity to certain odours varies from person to person). Thus, if you find you dislike the green bell pepper-like aroma in one glass of Cabernet Sauvignon, please give some other Cabernet wines a chance in the future. The green aromas in your next glass may not smell so awful to you and you may even find it pleasing to smell the vegetal odours in a third bottle of Cabernet Sauvignon. The wine industry says: 'There are no great wines, only great bottles.' I dare say (perhaps less catchily): 'There are no bad aromas, only bad aromas in specific wines, for specific individuals.'

10
Friends or foes? Foxy, musty and gamey smells

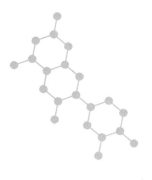

Aromas for arguments

Most of the aromas discussed in the previous two chapters give us pleasant sensations, or at least not revolting ones. In this chapter we will look at some aroma compounds that are perceived as unwelcome by most consumers; their presence is often seen as a flaw or faultiness in wine.* 'Foxy', 'musty' and 'gamey' are the descriptors addressed in this chapter. They may sound exotic since we don't often encounter them in our everyday lives and you will never know how you are going to react to their associated aroma compounds until you smell them. I still remember opening a bottle of red wine that reeked of gamey odours; its smell reminded me of the nasty stench of dead raccoon (yes, I've experienced that too), while a friend kept on sipping and enjoying the bottle for hours. As we've established well at this point, personal preference is everywhere in wine tasting experiences – and this was an extreme example of this phenomenon in action.

Perhaps I could have tolerated that red wine, had I forced myself to continue smelling it while debating with that friend. As my nose became used to the revolting animal smell, I might have

---

* For a comprehensive explanation of wine faults, please see *Flawless: Understanding Faults in Wine* by Jamie Goode.

become insensitive to it. This is called odour adaptation, which will be discussed later in this chapter.

The foxy aroma from grapes

Let's begin with the reason why only one grape species dominates the entire winemaking world, when there are more than sixty species out there. The vast majority of wines are made from the species called *Vitis vinifera*, which literally translates as 'the winemaking grapes'. This grape species contains thousands of varieties, such as Chardonnay, Sauvignon Blanc, Merlot, Cabernet Sauvignon and Pinot Noir, that we know of. The exact origin of *Vitis vinifera* is still debatable, but it is definitely a grape species that was popularized in Europe and Southwestern Asia. In North America and in East Asia, there are diverse grape species that grow locally. Unfortunately, these are not suitable for winemaking due to their unpleasant sensorial outcomes, which are often associated with bad and unusual smells.

As discussed in earlier chapters, fruits may produce off-aromas that deter animals for the purposes of self-defence. Even in *Vitis vinifera* varieties like Cabernet Sauvignon, methoxypyrazines are accumulated in the grape skins to drive away certain insect herbivores (see Chapter 9). Most grape species are more masterful than *Vitis vinifera* when it comes to producing unattractive odours. The most well-studied is the foxy aroma in the grape species *Vitis labrusca*. It is an aroma that reminds people of wet earth, the fur of a wild fox (if you have ever smelled it) and muskiness. But the best reference is to taste a fermented wine made from *labrusca* grapes. In Latin, *labrusca* means wild, which suggests the aromas that people may associate with wild animals like foxes.

Chemically, the foxy aroma comes from methyl anthranilate which contains carbon, hydrogen, oxygen and nitrogen. In Chapter 9, we explored the nitrogen-containing pyrazine compounds. In our daily life we tend to be driven away by ammonia, which consists of nitrogen and hydrogen. It seems that most of the volatile nitrogen-containing compounds are perceived negatively by our noses, as many such compounds are found in waste gases produced by organic living beings.

Concord is perhaps the most well-known grape variety within the *Vitis labrusca* species. We can find this variety almost anywhere in North America due to a brand called Welch's. The Welch's company produces juice, jam, jelly and many other consumable goods from the Concord grape. You may wonder why people like these products if they contain methyl anthranilate. This is because the compound, when found in non-alcoholic forms such as juice, is not perceived as foxy and the sweet taste from sugar masks the foxy smell further. This is another reason why most non-*vinifera* American and Asian grape species are consumed as fresh table grapes or in forms in which they have not been fermented.

It is worth mentioning that the most planted grape variety in the world is the Kyoho grape. This grape was created by a Japanese breeder in the 1930s via the crossing of a *Vitis vinifera* variety and a *Vitis labrusca* variety. Interestingly, Kyoho inherited more of its aroma blueprint from the *Vitis vinifera* side, so the methyl anthranilate compound is non-detectable in this grape. However, some foxy aromas are still obvious in wines made from it. Therefore, Kyoho is not a suitable variety for winemaking on a commercial scale.

Despite the many varieties of grape that give off unpleasant aromas, we still have a wide range of options when choosing the grape varieties suitable for winemaking. But the next aroma compound, unfortunately, is harder to avoid and has been an issue in the wine industry since its inception.

The obnoxious 'cork taint'

When a space has been dominated by microorganisms, distinct aromas are created by them. Some of them are desirable, such as the yeasts that occupy grape juice and turn it into wine with charming smells. On the other hand, yeasts and bacteria can create compounds that destroy the sensorial attributes that winemakers intended to showcase. The most notable and notorious chemical compound is 2,4,6-trichloroanisole (TCA), which costs the wine and cork industries millions every year in terms of the losses it causes and the investment spent to mitigate the problem.

Cork taint is the common name describing corks and wines that have been tainted by the TCA compound. This compound is created by several reaction steps in woody materials such as natural corks. First of all, natural corks are made from the bark of a particular oak tree species called *Quercus suber* (which is different from the oak species used for barrel production). The bark contains many phenolic compounds. When chlorine or chlorides in water come into contact with the cork, a simple phenolic compound can be turned into an intermediate compound called trichlorophenol. Some airborne microbes (fungi and/or bacteria) turn the trichlorophenol into 2,4,6-trichloroanisole (TCA). This very last step makes the not-so-volatile trichlorophenol become the extremely volatile TCA, causing people to smell the cork taint even from far away.

To summarize, the essential components in the formation of TCA are phenolics, chlorine and certain microbes. The first two are almost unavoidable as natural cork is filled with phenolics and the water we use almost always contains a tiny amount of chlorine. This means that even if cork manufacturers send out corks without any TCA contamination, there is a chance that some microbes in the air will create TCA later on. If natural corks, or wine bottles

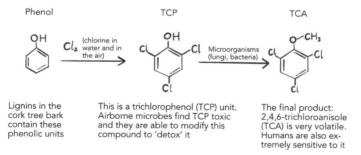

Figure 20: The formation of TCA

sealed with such corks, are stored in a humid environment this encourages the growth of many different microorganisms and there is a higher chance of having TCA issues. The elimination of TCA contamination is almost impossible to guarantee even in dry conditions, however, as there is always a chance of contamination simply from contact with the air. Thankfully, the chances of having a cork taint problem are much lower nowadays, due to more advanced methodologies and quality control (such as using people with an acute sense of smell to screen individual corks, or through analytical techniques).

Another key reason why TCA is so notorious is its extremely low (general) range of sensory threshold. Like the 3-isobutyl-2-methoxypyrazine (IBMP) discussed in Chapter 9, TCA is detected by many people at the parts per trillion level. The formation of a minuscule amount of TCA will ruin an entire bottle of wine. You might say that it is a punishment from Mother Nature for taking a bottle of wine too seriously. A bottle of the most expensive red Burgundy wine from the 1970s costs somewhere between US$10,000 and US$60,000 per bottle. A billionaire may own a bottle to open it on a special occasion, only to find upon opening it that the wine is corked. The bottle will probably still be consumed that day, because the taste of the wine is perhaps not the most important part of the experience anymore. 'Label-drinking' is everything for that bottle.

People may use different terms to describe the smell of TCA. Common descriptors include musty, mouldy, wet cardboard or a bag of baby carrots (which smells almost exactly like cork taint due to the favourable environment for incubating TCA in these packaged vegetables). But it is not just about the odour of the compound itself. TCA can also mask the desirable aromas in wine, especially the fruity and floral smells. The exact mechanism is not well-studied, but the explanation is simple enough: TCA is one of the warning signs of microbial spoilage and our nose has evolved to notice such odours of danger without being able to enjoy the other pleasant smells in the ambient environment.

TCA is not the only compound that causes cork taint. The other three well-studied compounds are 2,4,6-tribromoanisole (TBA), 2,3,4,6-tetrachloroanisole (TeCA) and pentachloroanisole (PCA). All of these, including TCA, are categorized as haloanisole compounds. But TCA is the major contaminant, as it is much more readily formed under normal conditions than the other haloanisole compounds.

Other microbial taints

There are other compounds produced by microorganisms that can taint a wine. Some famous microbial taints are associated with the gamey, animal-like smells in nature. The logic behind such association is that quite a few odours that we smell from the skins and furs of animals are caused by microbial activity. For example, our sweat as produced does not have any aroma in the instant the glands in our skin produce it. But as soon as bacteria in the air come into contact with the sweat, they produce volatile compounds which result in the body odour that our noses can perceive. The two well-studied microbial taints that cause the sweaty horse-like odours are *Brettanomyces* (which is the name of a group of yeasts) and mousy taint (which is the sensory outcome of a type of wine fault).

Brettanomyces

The name *Brettanomyces* refers to a genus of yeasts. Quite a few yeast species within this genus tend to produce compounds with distinct aromas. The most famous compounds are 4-ethylphenol (4-EP), which smells like a mixture of sticking plasters and a farmyard where horses are kept, as well as 4-ethylguaiacol (4-EG) which smells like bacon, cloves and smoke. In wine, the presence of 4-EP and/or 4-EG produced by *Brettanomyces* yeasts is often treated as faulty, as some people perceive such odours as unpleasant. In the wine industry, this fault is informally called 'Bretty'.

However, some consumers are not opposed to the Bretty smells and consider those aromas an additional layer of complexity

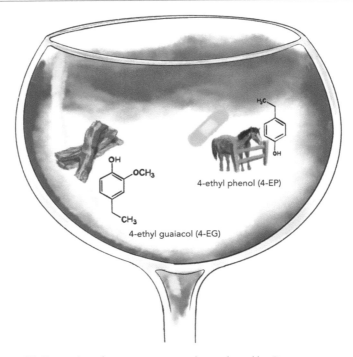

Figure 21: Examples of aroma compounds produced by Brettanomyces yeasts

in the wine. This observation is even more obvious in the beer industry, which may not consider Bretty smells as off-odours at all. During the brewing of certain Belgian ales, the activity of *Brettanomyces* yeasts is encouraged and the abundance of 4-EP and 4-EG in those ales is viewed as their signature aromas without any negativity. *Brettanomyces* can also produce volatile fatty acids like isovaleric acid and isobutyric acid which contribute to sweaty, cheesy and rancid smells. However, due to individual variabilities, those volatile acids may or may not be detectable by a particular person; these genetic differences will be discussed in the next chapter.

Mousy taint

As the name suggests, this off-odour smells like caged mice (or sometimes rotting meat). Although this type of fault can be caused by chemical reactions, it is most commonly produced by quite a few strains of lactic acid bacteria and certain species of *Brettanomyces*. The resulting compounds that contribute to the mousy smell are mainly those containing a pyrroline group (which looks like a ring structure with a nitrogen (N) element in it).

The sensory aspects of these compounds are complex, not only because sensitivity towards them varies significantly from person to person, but also due to the fact that they are not quite volatile at low pH. This means that wines with such taint may not show obvious mousy odours when we smell them (orthonasally), but once we taste the wines and our saliva neutralizes the acids to raise the pH in wine, the related compounds become volatile and can be detected via our retronasal olfaction.

'Reduction'

Here, the term reduction indicates a type of fault in wine. The meaning of reductive aromas or reductive winemaking will be explained in Chapter 12 but to put it simply, the majority of yeast species are capable of producing specific sulphides (sulphur-containing compounds explained in Chapter 8 and Chapter 9) which give stinky aromas reminiscent of rotten eggs, rotten cabbage and old onions.

Chemically, the sulphide compounds have a reductive nature. If the sulphides produced are low in concentration, their aromas are not perceived as negative and may even become part of the aroma complexity to some people. Also, as we will discover later in this chapter, low levels of sulphides in wine can dissipate by aeration in the bottle or in a glass. In this case, yeasts have not yet produced enough reductive sulphides to cause faultiness in wine, hence the quotation marks on the term 'reduction' in the title of this section. But in high concentrations, certain sulphides will exude overwhelmingly pungent odours that can last for a long time. In that instance, we say the wine is tainted by too much reduction.

It is worth noting that the acceptance of so-called microbial taints or flaws in wine depends very much on the food and beverages that people are accustomed to. Microbial activities naturally happen and bring more aromatic complexity. For example, wines influenced by certain strains of the *Brettanomyces* yeasts and bacteria like *Pediococcus* can smell similar to Bretty beer, pungent kimchi and foul-smelling (to many) fermented bamboo shoots, which are all popular in specific countries. With careful surveillance and control, yeasts and bacteria help food become more stable and nutritious, regardless of whether the food tastes delicious or unappealing to you. Thus, the longer the culinary history of certain fermented food products, the easier it is for people in those cultures to accept a wide range of aromas created by microorganisms.

Odour adaptation

Since any living thing is constantly adapting to anything that surrounds it, all of our senses have the tendency to get used to the same stimulus over time. The sense of smell, in particular, is rather quick in terms of adapting to a new odour in the ambient environment as we can easily observe. For instance, upon entering a smelly bathroom, we are met by an intense and unpleasant odour. After a few minutes, however, our nose gradually comes to tolerate the bad smells, despite the fact that the concentration of those odour chemicals remains at about the same level. Our odour adaptation abilities can be surprisingly strong. It has been observed that after exposure to specific aroma compounds, our reduced sensitivity to such compounds can last for two weeks, unless quickly replaced by other odours that help 'refresh' our sense of smell.

The basis of odour adaptation at the cellular, neurone and molecular levels has been well-studied. This is due to the economic and social significance of understanding odour adaptation. Not being able to detect toxic volatile compounds due to adaptation to them can put us in great danger, for example in scenarios such as getting used to the smell of poisonous gas in a chemical laboratory or to hazardous paint fumes in a newly painted indoor space. To give a commercial example, the mechanisms of many air fresheners are based on sensory perception rather than chemistry. The chemicals coming out of a can of air freshener do not react directly with unpleasant aroma compounds in the air. Instead, the spray of air freshener acts on our nose to dilute and mask our perception of the unpleasant smells.

On the other hand, because of odour adaptation, we can experience olfactory 'fatigue' from pleasant odours and sometimes need to find ways of refreshing our sense of smell. We can drink water to rinse and cleanse the palate, but there is no convenient agent that acts as a 'cleanser' for the nose. Anecdotally, coffee beans have been used as an olfactory cleanser, but studies show contradicting results in terms of whether or not the beans have refreshing, cleansing properties. There is no magic pill for rebooting our olfactory system to receive new odours except for resting our noses and bodies, surrounded by fresh air. Therefore, when evaluating any products that rely on the sense of taste and smell, judges must be cautious about how fatigued by familiar tastes and smells they may have become. Many wine competitions nowadays reduce the number of wines for experts to judge each day, in order to reduce the negative effect of odour adaptation.

Another important consideration is to diversify the style of wine between the groups. Evaluating solely Chardonnay wines for the entire day will cause severe olfactory fatigue. The solution is to let the wine judges taste a flight of Chardonnay, then swap to a different matrix such as a group of Pinot Noir wines. In our daily enjoyment of wines, trying out various kinds of wines is not only fun but also can reduce the negative effect of odour adaptation.

Chapter 5 explained the complications of taste interactions when it comes to food and wine pairing. What about the pairing of aromas between food and wine? The answer is that it is even more difficult to predict the aroma interactions. We may hear from chefs, sommeliers and gourmets that wines with gamey aromas should pair with meat that exhibits the same type of smell. Yet what if our nose adapts to the animal-like smell in wine and subsequently perceives less of the aroma complexity from the meatiness in food? Perhaps contrasting odours in food and wine give a better pairing experience, with a constant flow of new aroma compounds. Pairing the same aroma in food and in wine may not give us an enhanced experience, just like matchmaking: two people with the same personality may not work together when it comes to a relationship.

Does wine need to breathe?

You may often hear the saying that certain wines need to breathe. Indeed, some wines do not fully express their sensory characteristics unless we aerate them, such as by going through the decanting process. But do we need to decant a bottle of wine to let it breathe? The simple answer is the answer to nearly every scientific question: 'It depends.' There are two major theories behind the reasons for decanting a wine.

One is that, for wines with high levels of tannins, aeration can make the tannins taste smoother due to the oxidation and polymerization of tannins, anthocyanins and some other phenolic compounds (see Chapter 1 and Chapter 7). Chemically, this is not quite correct. In order for there to be significant sensory impact from the structural change of tannins, the time required is at least two or three days with full exposure to air at room temperature. Therefore, in terms of phenolic compounds, while aeration can quickly promote changes such as browning in white wines (see Chapter 2), altering the core compositions of tannins happens slowly. Only long-term ageing may significantly enhance the mouthfeel of tannins to reward patience.

However, some people have had experiences of perceiving a more rounded mouthfeel after decanting a bottle of red wine and letting it 'sit' for a while. There can be three reasons behind such observations:

- First, those individuals might have taken the wine bottle out of a cool cellar or wine fridge right before opening it. As the temperature of a bottle rises to room temperature, the mouthfeel of the wine (especially the tannins in the wine) tends to become smoother.
- Another reason is related to preconceptions (discussed in Chapter 3). If a person expects the wine to taste better after letting it sit, their brain may tell them it tastes better even if nothing has happened chemically or physiologically.
- The third reason is to do with sensory adaptation. When people revisit the wine after decanting it, they might have already had some other drinks or salty food, making their palate less sensitive to the astringency of tannins.

These scenarios are not measurable but are worth consideration.

Another theory of why some wines need to breathe is that the major aroma profile in wine can be hidden by other types of odours. This is scientifically more valid. For example, sulphur-containing aroma compounds like sulphides and sulphites (see p. 184) are highly volatile, and most people are very sensitive to these compounds. Consequently, when we sniff a glass of wine, the sulphides and sulphites can quickly 'grab' our olfactory receptors, getting in before the other aroma compounds and effectively concealing them.

Chemically, as the wine breathes, the sulphur-containing compounds can be oxidized rapidly, losing their aroma properties. Sensorially, odour adaptation has a role to play. Just as our noses can adapt to the smelly odours in a restroom, some low levels of off-odours in wine become weak after several sniffs. As the noise from the sulphides and sulphites fades away, we seem to be able to detect the 'true' aromas behind a glass of wine.

Final thoughts

So many aroma compounds can make a wine smell unusual: methyl anthranilate for the foxy smell, TCA for the mouldy smell, 4-EP for the farmyard smell, isovaleric acid for the rancid smell and hundreds of other sulphur- or nitrogen-containing compounds for varied odours. As we saw in this chapter, most of those smells result from microbial activity, which can be good or bad for the overall sensory profile of a wine.

As consumers, we choose the wine that we would like to try. But don't forget that we can also choose the ways that we drink the wine. Due to odour adaptation, the smell of a wine may change as time passes. A wine will also smell different if our nose is preoccupied by other prominent aroma compounds. So don't be restricted by the so-called rules when enjoying a glass of wine. If a red wine has off-odours that are unacceptable to you, heat it with sugar and spices to make a mulled wine. The heat can burn off the unpleasant smells and the spices of your choice can add pleasing aroma complexity (we will discuss spice aromas in the next chapter). As Charles Darwin said: 'It is not the strongest or the most intelligent who will survive, but those who can best manage change.'

11
Spicy, petrol and woody complexity

Spice it up

Countries with well-established culinary cultures have long known the charm of the 'not so green' parts of a plant. The seeds, roots, bark and fruit of certain plants have been used to craft the aromatically intense materials often used for cooking: spices. In countries such as India, spice-rich foods such as curry have become emblematic of the nation.

Most grapes possess few of the aroma compounds that give spicy smells. The exceptions include the black pepper-like smell, which we will look at below. Then we will examine the petrol-like aroma that frequently prompts questions and sparks debates in the wine industry. The main focus of the chemical terroir in this chapter is the aromas from oak, which serve as the seasoning in the production of wine. In the same way that spices are added during cooking, most spicy aromas in wine come from external sources that are introduced during the winemaking process. With a few exceptions, oak is the only material that is legally allowed to contribute extra aromas in wines. These oak-derived aromas are mostly spicy and woody to our noses.

In terms of the sensorial terroir, this chapter navigates through the nuances of specific anosmia, which is often the key barrier in terms of communicating what we smell in daily life. We will discover that in wine, a few aroma compounds have already been identified as distinguishing markers, separating individuals who can readily detect them from those who cannot.

From the grape: pepper spice

Pyrazines, as discussed in the last chapter, exhibit green bell pepper-like aromas. But these types of peppery aromas are usually more closely associated with green and vegetal smells than the aromas of the spices we keep in our kitchens. If I asked you to think of peppery aromas, your mind would probably conjure up an image of black or white pepper or peppercorns. This is because the enticing aromas of black and white pepper have become ubiquitous. The story of the black pepper-like smell in wine is fascinating.

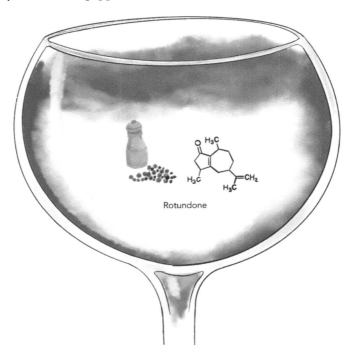

Rotundone

Figure 22: The aroma marker of black pepper. Rotundone is the key aroma compound responsible for the signature smell of black pepper. Wines made from certain grape varieties like Syrah/Shiraz also contain rotundone and so can show black pepper aromas.

In 2008, a group of researchers in Australia identified the key chemical compound that contributes to the black pepper aroma in wines made from the Shiraz grape variety. This compound is called rotundone, a type of terpenoid that is widely present in herbs and black or white peppercorns.* Interestingly, rotundone was not recognized as the most important aroma contributor in actual black and white peppercorns until those Australian researchers correlated the chemical and sensorial aspects between Shiraz wines and peppercorns. In black or white peppercorns, the concentration of rotundone is about 10,000 times higher than the level in Shiraz wines.

You may wonder why the flavour chemistry world did not identify the sensorial significance of this compound prior to its discovery in the wine world. The answer is simple: it is only in recent decades that the analytical machines and the sensory methodologies have become advanced enough to recognize those aroma compounds that, while extremely low in concentration, have a significant impact upon the aroma. The chemical properties of many aroma compounds were identified a long time ago, but the research world did not have the tools for correlating the chemical and sensorial matrices back then.

If anyone finds the experience of detecting black pepper aromas in wine unusual, the next aroma we will discuss will prove even more exceptional.

TDN (1,1,6-trimethyl-1,2-dihydronaphthalene): an oddball

As we noted in Chapter 8, terpenoids can smell unusual and even a bit unnatural to us. In wine, one terpenoid has been a hot topic for many years, purely because of its strange aroma. This terpenoid has a long and unappealing chemical name – 1,1,6-trimethyl-1,2-dihydronaphthalene – which we mercifully shorten to TDN.

---

* The other grape variety that is famous for containing high levels of rotundone is Grüner Veltliner, Austria's most famous white variety.

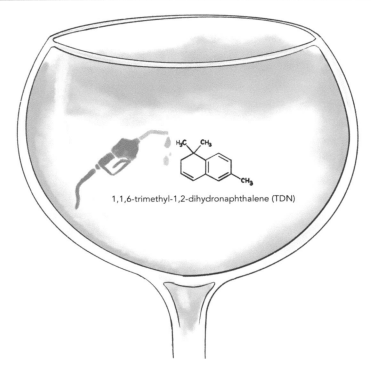

1,1,6-trimethyl-1,2-dihydronaphthalene (TDN)

Figure 23: The aroma marker of 'petrol' in wine. The compound TDN gives wines a distinct petrol-like aroma. Wines made from the Riesling grape variety often express this aroma character.

To some people, the smell of the chemical resembles petrol or kerosene, making it unattractive in aroma as well as in name.* To other consumers, though, TDN gives a desirable smell that adds complexity to a wine.

The best wines to demonstrate the effects of TDN are those made from the Riesling grape. When the growing conditions are sunny and warm, or after ageing for a period of time, Riesling wines tend to develop TDN. Because of the chemical's high

---

* Chemically, this is hardly surprising as both terpenoids and petrol are based on carbons. Nowadays, certain biofuels are synthesized from terpenoid units.

volatility, people can detect its aromas instantly. It is a memorable aroma compound that helps consumers recognize and identify Riesling wines easily.

After that diversion on the unique petrol smell of TDN, it is time to redirect our focus to the spicy realm. The topic is about one of the most captivating companions of wine: oak.

Oak and wine: a love story

Wooden vessels have been an integral part of human civilization for centuries. Skilled craftsmen can turn any wood into tools that have seemingly magical properties. One of the most amazing cultures that coexists with winemaking history is the art of oak barrel making. Based on research evidence, some scientists even speculate that the principal yeast species for fermentation – *Saccharomyces cerevisiae* – were originally sourced from oak trees, which the microorganisms inhabited. By some strange twist of Mother Nature, certain oak species possess everything we desire in order to elevate wines to another level. Barrels made from oak tend to be more water-tight than those made from other types of wood and the aromas of oak marry with the flavour profile of wines seamlessly. Oak could be the subject of an entire book and it is still not well-understood by the wine and spirit industries.

In this book we focus on the chemical and sensorial aspects of the love story between oak and wine. The following discussions break down the key chemical markers for oak aroma compounds and their sensorial impact on wine tasting. (Note that the impact of oak tannins on the astringency sensation perceived by our tongue has been discussed in Chapter 7.)

Furfural: toasted almond, toasted bread, caramel-like aroma

Let's start with the compound that gives the generic smell of toasted wood and nutty materials: furfural. Furfural is derived from hemicellulose, which is one of the most abundant carbohydrate compounds in the woody parts of plants, forming their supporting structures. It is a type of aldehyde and also features in the oxidative realm, which we will look at in Chapter 12. With a heating process such as toasting, a significant amount of furfural will be created from hemicellulose.

An oak barrel maker is called a cooper and a manufacturer of oak products is called a cooperage. When a cooper makes an oak barrel, toasting is the essential heating process for bending the wooden staves to shape the barrel. During the toasting process, furfural and many other aroma compounds will be created. Even with oak barrel alternatives like oak staves and oak chips, methods of heating are almost always applied to generate the desired aromas.

Any woody materials or items containing hemicellulose-like carbohydrates that have been toasted may emit a somewhat similar smell: toasted almond, toasted bread and caramel all smell a bit like toasted oak. One of the top manufacturers of oats, the Quaker Oats Company, was an early mass-producer of furfural chemical extract, which they made using the left-over oat hulls. Since then, the commercial use of furfural has surged and it is used in a wide range of products as a flavouring compound. Furfural is one of the key compounds that link these items together in our brain when we smell them.

Yet, there is hardly any item that has furfural as the predominant aroma compound. This is why we tend to use a combination of aroma descriptors (including almond, toasted bread and caramel, as described earlier) to nail down the smell of furfural.

In wine, furfural can contribute to a wide range of aromas as well, depending on the other aromas in the matrix. Recent studies have shown the interactions between the compounds in wine and those in oak. For example, 2-furfurylthiol is the major compound responsible for the typical smell of coffee. When the thiol (sulphide) compounds (see Chapter 8 and Chapter 12) in wine react with the furfural compounds in oak, 2-furfurylthiol can be produced, hence the coffee-like aromas in wine. Some South African producers make oaked wines from Pinotage that smell almost exactly like coffee without any external flavourings added, a great example of the perfect marriage between wine and oak.

Lactone: coconut-like aroma

In tropical areas where mosquitoes and flies are extremely prevalent, plants tend to produce quite a few volatile compounds to repel those insects. The coconut fruit is filled with different types of lactone compounds that act as repellents. Therefore, any material that contains enough lactone compounds can remind us of the smell of coconut. Chemically, lactone is a special type of ester that forms a ring-like structure. While the chemical is unattractive to insect herbivores, most people find the aromas of lactone pleasant; as evidenced by the popularity of coconut-flavoured drinks and coconut-based cuisines.

The types of lactones found in oak are commonly (and predictably) called oak lactones. They are also called whisky lactones as the majority of aromas in whisky come from the oak. Unlike most other aroma compounds, oak lactones can be considered the natural aroma compounds of oak as they are abundant in the raw wood. In fact, the toasting of oak tends to decrease the concentration of oak lactones.

Just as different grape varieties produce distinct aroma profiles, the ability to synthesize lactones is highly dependent on the oak species. The American oak species (*Quercus alba*) always produces much more oak lactone than European or Asian oak species (e.g. *Quercus petraea*, which are widely grown in Europe). This is why at the same toasting level, wines influenced by the American oak express more sweet coconut-like aromas than those that have had contact with French oak.

Guaiacol: smoky aroma

Guaiacol is probably the most familiar scent of danger that we can recognize, since it is the one emitted when wood is burnt. It is one of the direct signals in nature that tell animals to run away from fire. But humans have culinary cultures that rely on (an appropriate amount of) fire and heat. With experience, we have developed methods of controlling the level of guaiacol in barbecue and other smoked dishes. A desirable level of guaiacol will make us hungry, while too much guaiacol warns us to turn on the ventilation system.

Chemists have discovered other magical qualities of guaiacol. For instance, the chemical synthesis of vanillin and eugenol (see p. 169) involves guaiacol as a key component. In other words, guaiacol is one of the precursors in synthesizing vanillin and eugenol. The raw chemical materials of guaiacol and glyoxylic acid are part of a two-step process that yields the majority of vanillin that is widely used around the globe.

A hot topic in the wine industry nowadays is the smoke taint of grapes and wines caused by the more frequent bushfire events in recent decades. When a bushfire event occurs, grapes exposed to smoke in the air can absorb a high concentration of smoky compounds. As a result, the wines produced may exhibit unbearably ashtray-like odours. Guaiacol, along with some other compounds that cause the smoky taint in wine, are the key chemical markers to winemakers and researchers. Considerable funds have been devoted to the research of smoke taint, in order to analyse the damage caused by bushfires more precisely and to find efficient ways to mitigate or to remove such taint or fault in the aftermath of natural disasters.

In an oak cooperage, smoky compounds can be well-controlled, with higher levels of toasting generating more guaiacol and other related compounds. A hint of smoke is desired in certain wine styles and adds to the complex aroma profile of oak.

Vanillin: vanilla-like aroma

Although natural vanilla beans have hundreds of aroma compounds, vanillin is the major contributor to the smell of vanilla. It seems that vanillin was born to become a key flavouring agent for human beings. The food and beverage industries are immersed in vanilla-flavoured products and they always sell very well. For unknown reasons, we are drawn to vanilla-like aromas.

In oak, the majority of vanillin compounds are derived from another key structural compound of plants called lignin. When heated, lignin can be broken down into many volatile compounds including vanillin. In fact, up until the 1980s, most vanillin extracts for commercial and industrial use came from the lignin-containing waste produced in manufacturing papers. Even today, 15 per cent of the vanillin supplies in the world are derived from lignin from such processes. The other 85 per cent use guaiacol as one of the precursors, as discussed above.

Due to its popularity, the sensory effects of vanillin have been widely studied in various food and beverage products. Based on the many studies available, we can only draw one conclusion: the perception of vanilla-like aroma is highly dependent upon the matrix of the medium flavoured by vanillin. For example, vanillin in water smells very different from the same concentration of vanillin in milk.

This is why food and beverage manufacturers continue to perform sensory tests on the target vanilla-like aromas. If the sugar and milk compositions are changed in an ice cream, the amount of vanillin used will likely need to be changed too or the consumer will notice a difference. Similarly, the same oak recipe cannot be applied to two types of wines. With the same amount of vanillin influence from the same oak barrels, the smell of vanilla will turn out to be different in its intensity and its nature between a Chardonnay from the cool climate Chablis region and, say, a Chardonnay from a warm part of California.

Eugenol: clove-like aroma

Almost all the chemical compounds discussed in this chapter have a fascinating historical background. This is the same for eugenol, which has long been an important compound for human civilization. For centuries, clove has been used in traditional Chinese medicine to relieve tooth pain. The essential compound in cloves that relieves pain is eugenol. In 1834, a chemist extracted eugenol from clove oil and this compound became the most popular dentistry tool for the relief of tooth pain. On top of that, the aroma of eugenol is attractive to many of us. Quite a few delicious cuisines are flavoured with cloves. There is no doubt that eugenol is the key aroma contributor to the clove-like smell: clove will not smell like clove without eugenol. Some people have noted a relationship between the smell of clove and those of cinnamon and basil – they also contain a certain amount of eugenol.

Eugenol is another chemical that can be used as a precursor of vanillin. Therefore, we can see that guaiacol, eugenol and vanillin are essentially close relatives in their chemical structure. Also, like vanillin, eugenol is one of the many compounds that originates from the breakdown of lignin in wood. As the toasting level of oak gets higher, the concentration of eugenol increases. At high concentrations, the eugenol odour can be pretty pungent – as we mentioned earlier, the pronounced smoky smell of the guaiacol

Figure 24: The aroma markers of oak

compound is the precursor of eugenol, so both compounds may act on our sense of smell in similar ways. Humans are naturally sensitive to such compounds, as they can act as a signal for toxicity, or potentially harmful situations. The reduction of tooth pain by eugenol is not because the compound kills any bacterial infection, but due to the suppression of nerves. Therefore, a high concentration of eugenol is not beneficial to our health and it was proven that an overdose of eugenol can cause liver toxicity.

In food and beverages, the concentration of eugenol is in the absolute safe zone. The sweet spice-like aromas in wine infused by eugenol in oak contribute a desirable complexity to wines. When combined with the intrinsic spiciness of 'Rhône' grape varieties such as Syrah and Grenache, we get wines that are almost like tasting from a set of spice jars.

Other factors of oak

The oxidative effect of maturation in oak barrels will be discussed in the next chapter. Other than that, the aromas associated with oak can be influenced by many other factors, such as the growing environment of the oak, the seasoning of the wood, the various levels of toasting and heating and the shape and size of oak materials used. All of those considerations are worth further research and examination. Having discussed the chemical interactions between oak and wine, which give various wines their distinct aroma, let us move on to the sensory component of this chapter.

Specific anosmia

Specific anosmia is the inability to perceive a specific aroma compound, even though the sense of smell itself is not damaged. This is a common phenomenon rather than a disorder; most people have a normal smelling function, but likely have some aromas that they are either hypersensitive or insensitive to. When you consider that our noses can distinguish more than a trillion odours, it is hardly surprising that each person's odour receptors may respond differently to a group of specific volatile compounds. Research has also proven that specific anosmia is prevalent among humans.

The primary reason behind specific anosmia is genetic. In studies that aimed to profile the olfactory receptor genes, it is difficult to find any two individuals who possess the same genetic profile. These variabilities are often observed in human and food aroma interactions.

In some experiments, nearly half of the panellists could not perceive the black pepper-like aromas of rotundone at a concentration that was easily detected by the others. Observations like this frequently occur during the tasting of Syrah wines, which are often rich in rotundone. While some people can easily identify Syrah in a blind tasting due to their high level of sensitivity towards rotundone, others may find themselves wondering what exactly the smell of black pepper means in wine.

In any case, we should be aware of the different reasons behind specific anosmia. The first is the inability from birth to perceive specific aroma compounds. This means that the genes for detecting certain aromas either don't exist in some people, or have never been activated. Beta-ionone, which contributes to rose or violet-like aromas in wine (see p. 125), was undetectable to almost half of the panellists in a controlled experiment unless the concentration was increased to more than 500 times the level that could be detected easily by the other half of the participants.

Regardless of training, the ability to detect beta-ionone did not seem to improve or decrease among all the panellists. Since wines made from the Pinot Noir variety tend to be rich in beta-ionone, it is not uncommon to find rose, violet-like aroma descriptors in people's tasting notes. But to some consumers, wines made from Pinot Noir never smell floral. Rotundone, the pepper-like aroma compound, mentioned above, is indeed another great example of specific anosmia. Australian researchers discovered that 20 per cent of a group of panellists in an experiment could not detect the black pepper-like aroma even if the researchers spiked the Shiraz wines with at least a hundred times more concentrated rotundone compounds, while some others could smell the aroma at extremely low concentration.

Specific anosmia can also be associated with age. For example, androstenone was the first mammalian pheromone to be identified by scientists. It was observed that as people age, the sensitivity towards androstenone can decrease.

Another interesting observation is that specific anosmia can be 'cured' under certain conditions. Through training, some people can improve their ability to recognize aroma compounds like diacetyl, which contributes to a butter-like aroma in wine and beer. This is either because the constant exposure to this compound can awaken the olfactory genes for detecting diacetyl, or because the person could always detect the aroma, but was unable to connect the smell to its descriptor (butter-like) until after training on odour identification (as explored in Chapter 9).

People also react differently towards those aroma compounds in wine that can indicate faultiness. Chapter 10 explained that certain compounds that *Brettanomyces* yeasts produce can cause

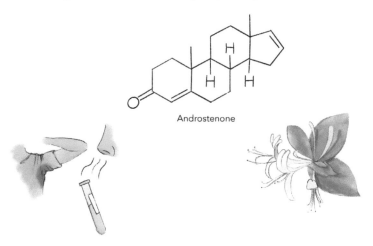

Androstenone

Figure 25: Different aroma perceptions of androstenone. To some people androstenone has an unpleasant smell, such as sweat or urine; to others, androstenone has a pleasant smell, such as wildflowers.

unpleasant smells, but these are not perceived as unpleasant by everyone. Isovaleric acid and isobutyric acid produced by *Brettanomyces* are well-studied compounds that bring to mind extremely pungent sweat and rancid cheese to some people because they have a genetically heightened sensitivity towards isovaleric and isobutyric acid. This is one of the reasons why certain tasters cannot tolerate any wine that contains even a tiny trace of isovaleric and isobutyric acid generated by *Brettanomyces*, while due to their anosmia towards the compound others may debate whether the same wine is faulty or not.

Lastly, it should be noted that our differences in perception when it comes to the same aroma compound are not only about being able to detect the compound or not. In some cases, the same compound suggests different types of smells to different people. For instance, in the case of androstenone, the compound is perceived by some individuals negatively as redolent of sweat and urine, yet other people detect it as sweet and floral and find it quite pleasant.

Final thoughts

This chapter provided more descriptors to ascribe to our glasses, such as the black pepper aroma from rotundone and the petrol aroma from TDN. Oak, by itself, contributes to several new groups of chemical compounds that attach a full range of spicy, smoky and nutty notes to a wine. Whichever types of aromas you prefer, remember that you might perceive some aromas which are non-existent, or perhaps even completely different, to those sensed by another person.

After considering all the spices in this chapter, we are reaching the final stage in our chemical and sensorial journey of wine tasting. In the next chapter, we will bring the book to a near-close by looking at the aromas created by the marvellous reactions of oxidation and reduction.

12
The world of oxidation and reduction

Yin and yang

The concept of yin and yang is rooted in ancient Chinese philosophy and serves as the foundation for traditional Chinese medicine. Supporters of this medical tradition emphasize its efficacy, pointing to thousands of years of successful treatments. In contrast, sceptics label it pseudoscience, citing a lack of comprehensive scientific validation and elusive methodologies. Nevertheless, recent research has uncovered correlations between the yin and yang balance in specific traditional Chinese medicines and the scientific principle of oxidation and antioxidation (reduction) balance.

My intention is not to ignite a debate on the credibility of traditional Chinese medicine. Instead, I am indulging myself by using the yin and yang philosophy to conceptualize oxidation and reduction, which is the focal point of this chapter. When delving into the study of reduction in chemistry, I envisage the yin elements, embodying counteractive properties like darkness, night, winter and the moon. On the other hand, when exploring oxidation, I see it as yang, representing proactive attributes such as brightness, day, summer and the sun. As we will learn in this chapter, despite being opposing forces, like yin and yang, reduction and oxidation coexist, interchanging electrons to achieve chemical balance.

This chapter will illustrate how the dynamics of chemical reactions shape the intricate cosmos of wine aromas. The yang aspect delves into oxidative aroma compounds, a frequently discussed topic in the context of wine oxidation. On the yin side, we explore reductive aroma compounds, a concept often inaccurately expressed in the wine trade. Through this exploration, we will aim to bring clarity to the meaning of reduction in the context of wine. At the end of this chapter, we provide context to underscore the extraordinary power of our sense of smell.

The world of oxidation

When it comes to oxidation, it might be easy to envisage the browning of apple slices and white wines (as we saw in Chapter 2). But what about the smells associated with oxidation? We will begin with the familiar process of cooking, which involves expedited oxidation (due to heat) and then explore the aromas as a result of oxidation in wine.

Oxidation in cooking

With exposure to oxygen or any other oxidants, almost all organic matter will go through the process of oxidation. As heat is usually involved in cooking, the process of oxidation in food is accelerated. When vegetables, proteins, oils, sugar, spices and all kinds of other materials are heated, parts of the aromas arise from oxidative reactions and they generally smell pleasant and complex. Wine is never meant to be heated unless we make mulled wine or Glühwein. However, the oxidation process will still take place if oxygen is present, simply over a longer period of time.

Wine professionals consider certain aromas to be oxidative, while other odours are called reductive. Such categorization is mainly based on the chemical understanding of winemaking and wine ageing. With significant exposure to oxygen, a wine will develop volatile compounds like acetaldehydes and ketones which are indeed more oxidative in their chemical nature. On the other hand, with little oxygen influence during winemaking, wines tend to have compounds like sulphides which are chemically on the reductive spectrum and contribute to smells that are very different to the oxidative ones.

Chemical oxidation

Chemically, wine is prone to oxidation due to the abundance of phenolic compounds that we encountered earlier (check out the Phenolics Lessons in Chapters 1, 2 and 7). As we saw in Chapter 2, phenolics, oxygen and metal ions are the three elements that initiate the oxidation process in grape juice and in wine. In this chapter, we are going to examine the further consequences of oxidation. In other words, the oxidation products, especially those that contribute to the oxidative aromas, will be the focus of our discussion.

When phenolics react with oxygen (catalysed by metal ions), two major types of chemical products will form. One is the quinones, which tend to react with other compounds like sulphur dioxide or to polymerize and become large molecules like dark-coloured melanin (discussed in Chapter 2). The other is peroxides such as hydrogen peroxide, which possess strong oxidative power and can oxidize many other compounds in wine.

For example, all types of alcohols become aldehydes, which tend to have pronounced smells (caramel, nuts) associated with oxidation. But aldehydes cover a wide range of smells. While some floral scents come from terpenoids (see Chapter 8), phenylacetaldehyde also reminds us of flowers like roses and it is a common ingredient in perfume to amplify the floral fragrance. Acetaldehyde is even more important to wine lovers as it exists in almost all alcoholic beverages. At lower concentrations,

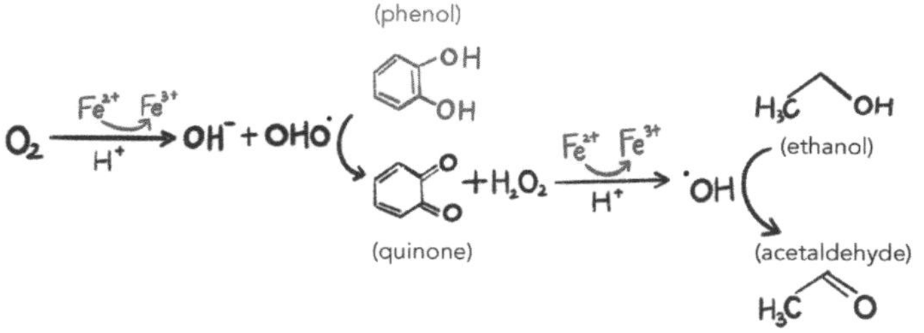

Figure 26: The road map of oxidation, simplified. A series of chemical reactions (you don't need to understand them) need to happen in order to oxidize ethanol to acetaldehyde. The key players in the oxidation reactions include iron (as Fe^{2+}) and phenolic compounds.

acetaldehyde smells somewhat fruity. If the concentration is high, it has a distinct rotten apple, Sherry-like odour. Here, the best reference is the fino style of Sherry. During the winemaking of fino Sherry, the base wine is made to have no residual sugar and with an alcohol level of around 15% abv. To survive in this liquid medium without sugar, some yeasts convert ethanol to acetaldehyde (instead of converting sugar into ethanol). As a result, acetaldehydes dominate the aromas to become the hallmark of the fino style of Sherry. In most other wines, such a high concentration of acetaldehyde would be considered a fault. Most people nowadays are not fond of such aromas, leading to a much lower market share for these historic Sherry wines.

Acetaldehyde is also known as ethanal, which is the oxidative/aldehyde form of ethanol. Other aldehydes such as octanal come from the higher alcohol compound of octanol. You have probably realized that the ending '-al' applies to the aldehyde and the ending '-ol' indicates the corresponding alcohol. Such chemical nomenclature not only demonstrates the chemical reactions of the oxidation of alcohols to aldehydes, but also indicates the change in the aromatic nature.

As discussed in Chapter 9, the compound 1-hexanol exhibits grass-like aromas. But as mentioned, cut grass shows a higher intensity of grassy smells because some 1-hexanol (ending in '-ol') is oxidized to the aldehyde hexanal (ending with '-al'),* which has much more pungent aromas. Sensorially, this is quite significant: we tend to have low thresholds to detect the smell of aldehydes. In other words, our nose is more sensitive to aldehydes as compared to the corresponding alcohols before being oxidized.

Unlike with cooking, which involves heat, the oxidation process in most wines tends to happen slowly. This is not only due to the lower temperature of wine storage, but also because of the antioxidative elements in wine. The most famous types of antioxidants in wine are phenolics and sulphur dioxide (SO_2).

---

* In Chapter 9, the aldehyde presented was *cis*-3-hexenal, which is an unsaturated isomer of hexanal

When the concentrations of both antioxidants are sufficient, only part of them and a small proportion of other organic compounds are oxidized slowly over time, resulting in many forms of oxidation products that contribute to just the right degree of complexity. Note that the antioxidants do not mean that they are immune to oxidation. It is quite the opposite: they are the sacrificial compounds that get oxidized first, so that the remaining compounds are protected from oxidation.

Only those wines with deliberate oxidation in winemaking result in the predominant oxidative characteristics. For example, oloroso Sherry and Vin Santo wines are deliberately exposed to lots of oxygen during production; the wines in barrel maturation are not fully filled, in order to provide lots of headspace for oxygen exposure. Hence, those wines are the best examples of understanding the oxidative aromas. In the next section, we are going to discover the quicker route to obtaining oxidation products with the aid of microorganisms like certain yeast species and bacteria.

Microbial oxidation

Chemical oxidation is not the only way in which wines can be oxidized. Some oxidation that occurs in wine is facilitated by microorganisms. During the metabolism of all organic matter, compounds with an oxidative nature, such as aldehydes and ketones, are the key intermediate metabolites. For example, the last step of alcoholic fermentation by yeasts is to convert acetaldehyde to ethanol. Therefore, it is common to find aldehydes and ketones created by yeasts and bacteria during winemaking. Even though those compounds are the product of microbial activity, their aromas remind us of oxidative food or beverages.

As discussed above, although the goal of yeasts is to create ethanol during fermentation, it is inevitable that they leave a small amount of acetaldehyde behind. A healthy alcoholic fermentation can always keep the acetaldehyde concentration to a minimum, but certain yeasts in harsh environments cannot yield ethanol efficiently and create many other by-products. The production of too much acetaldehyde by yeasts can be one of the undesired consequences when those microorganisms are stressed.

But how much is too much? In most wines, if there is an obvious smell of bruised apples due to acetaldehyde, we consider the wine to be faulty as it has been overly oxidized. A few wines, though, are glorified with the predominance of acetaldehyde, such as the fino-style Sherry, mentioned earlier. During the production of these Sherries, a layer of living yeasts called the *flor* (see p. 85) starts to develop on top of the wine. The *flor* yeasts are the same species as those conducting the alcoholic fermentation; however, they live under special conditions with oxygen above them and dry wines at specific alcohol levels underneath. In simple terms, the *flor* yeasts are floating on top of a wine with no sugar left in it. In order to survive, *flor* yeasts switch their diet to the consumption of alcohol and convert the alcohol to aldehydes. At this stage, this is probably the only way for the yeasts to metabolize and generate some energy to live. As a result, a handful of acetaldehydes from ethanol and other aldehydes from other alcohols are produced.

As mentioned, most people consider the Sherry-like, bruised apple aromas in other wines to be unappealing. But fino-style Sherry, along with other wines with the influence of *flor* yeasts (e.g. amontillado Sherry, vin jaune), are characterized by the tangy sensation they inherit from the aldehydes. Whether we like or dislike those wines tends to depend on preconceptions and personal experience. The aldehyde aromas should be appreciated in some Sherry wines, but the Sherry-like aromas suggest faultiness in most other wines. This is like our expectations of wine versus vinegar: a wine should not smell like vinegar and will only do so if there is a fault in the wine, or it was not properly sealed. Nobody would drink a wine that had a pungent vinegar aroma! However, if we have a bottle of vinegar that we expect to have that sour taste and smell, we're happy to put it on our food. On that note, let's examine the microbial oxidation that turns wine into vinegar.

Although ethanol can be converted to vinegar with different catalysing agents, the most efficient catalysts in nature are acetic acid bacteria. The name of the bacteria obviously suggests its *raison d'être* is to make the essential compound of vinegar – acetic acid. To produce vinegar, yeasts take the initiative of fermenting any solution with simple sugars into ethanol. Then, acetic acid bacteria convert ethanol to acetic acid. Therefore, the fate of grape juice is to eventually become vinegar, should it follow the order of Mother Nature, while winemakers try every means to prevent the alcohol from being transformed into vinegar.

Unlike other types of acids covered in Chapter 5, acetic acid is volatile, so it contributes to aromas on top of sour tastes. Sensorially, a low level of acetic acid in food can be favourable. While in many culinary cultures vinegar is used as a flavour enhancer in cooking or as a dipping sauce, in wine production, most winemakers try to avoid the development of acetic acid as it is hard to control the process once acetic acid bacteria start to spoil the wine.

One last point about acetic acid. If a wine had a high level of acetic acid, it would no longer be considered a wine. Therefore, almost all countries set legal limits on the maximum amount of acetic acid in wine. For example, in European Union countries, a red wine with more than 1.2 grams per litre of acetic acid cannot be sold to the market. Depending on the country and regulatory bodies, a liquid that contains 40 grams of acetic acid per litre or above is usually defined as vinegar. This leads to the puzzling consideration: how unfortunate it must be for a liquid to contain somewhere between 2 and 40 grams per litre of acetic acid, since it falls into neither the category of wine nor of vinegar. I wonder if Thomas Jefferson's bottle of Château Lafite 1787 had such levels of acetic acid when it was sold at auction in 1985 for US$156,450.

Complex oxidation process

Many of the compounds associated with wine oxidation have not been well-studied in the academic world. This is because the formation of some compounds takes many steps and not all of the intermediate products in the chemical reaction pathways can be captured by researchers. This chapter highlights a few exemplary compounds that are more familiar to the wine industry in terms of their chemical and sensorial nature.

Diacetyl is a famous compound, not only in the wine world, but also in the beer and food industries. The smell of diacetyl is very close to the aroma of butter, which makes it a beloved compound in many countries where butter is an important component in cooking. There are commercially successful Chardonnay wines that are driven by the buttery aroma of diacetyl. In winemaking, such an aroma can be controlled by encouraging the completion of malolactic conversion, which involves the lactic acid bacteria that convert the malic acid in wine into lactic acid. However, diacetyl is only a by-product during malolactic conversion. The key contributors are citric acid (as the precursor) and the specific strain of lactic acid bacteria involved. Therefore, not all wines that have been through the malolactic conversion show obvious buttery aromas. Diacetyl is regarded as a product of microbial oxidation due to its oxidative chemical nature.

Phenylacetaldehyde is a combination of a simple phenolic unit and an acetaldehyde unit. It was found in high concentrations in Sauternes wine samples. In pure water solutions, phenylacetaldehyde smells like honey as well as rose bush (a combination of green and floral smells). The formation of this compound comes from Strecker degradation, which is about converting an amino acid into an aldehyde. Aldehyde compounds have a more oxidative chemical nature, meaning they have a tendency to 'grab' electrons from compounds that tend to 'donate' electrons. In this case, a simple phenol gives electrons to an acetaldehyde to form phenylacetaldehyde.

Sotolon is a compound found in a clove-like herb called fenugreek. Nowadays, it is widely recognized in the world as a flavouring component in cooking. Its aromas can be variable: at low concentrations, sotolon gives off sweet smells of maple syrup, molasses and caramel; at high concentrations, sotolon reminds people of curry spice. We still do not quite understand how sotolon is formed in wine, but several types of wines are famous for having higher concentrations of this compound. Madeira wines, which contain high levels of sotolon, are not only oxidized but also heated during production, which makes researchers wonder if the deep oxidation accelerated by high-temperature conditions helps promote the formation of sotolon. Sotolon is also present in amontillado-style Sherry and vin jaune, which are influenced by the *flor* yeasts as well as oxidation during the maturation stage of winemaking. Therefore, both oxidative and microbial factors may offer more precursors for yielding sotolon.

Last but not least, grapes infected by the famous fungus known as 'noble rot' or *Botrytis*, used in making classic, premium sweet wines like Sauternes, Tokaji Aszú and Trockenbeerenauslese (covered in Chapter 4), can develop abundant sotolon compounds during production. Whatever the specific pathways that eventually create such aromatic compounds, oxidation seems to be essential.

The world of reduction

In chemistry, a compound with a reductive nature is prone to give an electron (or electrons) to a recipient and the recipients are those oxidative compounds. Therefore, some chemical analyses, including those that are occasionally used in winemaking, examine the redox potential (see p. 188) which is about a compound's tendency to acquire or to lose electrons. A pair of oxidative and reductive compounds are, as mentioned at the beginning of this chapter, almost like the yin and yang of nature; they coexist and are drawn to each other to create myriad dynamic reactions.

As we saw earlier in this chapter, the more oxidative compounds like aldehydes can vary greatly in terms of their aromas. This is the same for the reductive compounds that can contribute to smells. But the smells of reductive compounds are rarely similar to those in the oxidative world, which makes sense when we consider that our noses have adapted to distinguish the products from oxidation rather than those created in the absence of oxygen.

The main chemical group of reductive compounds in wine contains the element sulphur (S). We encountered quite a few sulphide compounds in Chapters 8 and 9, as plants, animals and inorganic beings produce a lot of them. In the wine field, there has been confusion and misinterpretation of two categories of sulphur-containing compounds: the sulphides and the sulphites. Below, we will look at both compounds and offer a comparison of the two.

Sulphide

When we say sulphide compounds, we refer to compounds that contain the ion of S^{2-}. This is a form of the element sulphur that is negatively charged, meaning it is reductive and prone to donate electrons to oxidants. The positively charged hydrogen ion (H^+, see p. 185) can react with S^{2-} to form specific compounds like hydrogen sulphide (H_2S).

In wine, any compound with a chemical group of -SH in it is regarded as reductive and known as a thiol compound. In Chapter 8, we saw the grapefruit-like aroma compound of 3-mercaptohexan-1-ol, which contains -SH in its chemical structure. Other than H_2S and thiols (any compound with -SH), there are also compounds with an S-S (sulphur-sulphur) bond within them, hence the name disulphide for those compounds. Disulphides are also included in the sulphide category as most of them are also associated with reductive aromas in wine.

Sulphite

Once we are clear about the chemical makeup of sulphides, it is easy to distinguish them from sulphites. Sulphite has the oxygen (O) element within its structure, in addition to sulphur. Typical examples of a sulphite include sulphur dioxide (SO_2) and the sulphite ion (SO_3^{2-}). Strictly speaking, the molecular form of SO_2 is not a sulphite but it is considered as such in the wine world for the sake of pragmatism. The chemical nature of sulphites is also reductive, but sulphites are treated more as preservatives rather than aroma donors by winemakers (even though a very high level

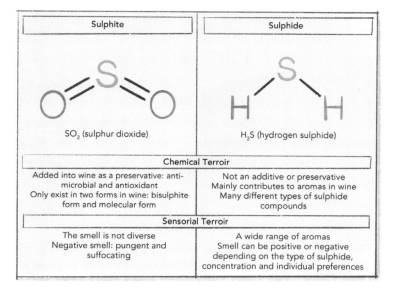

Sulphite	Sulphide
SO_2 (sulphur dioxide)	H_2S (hydrogen sulphide)
Chemical Terroir	
Added into wine as a preservative: anti-microbial and antioxidant Only exist in two forms in wine: bisulphite form and molecular form	Not an additive or preservative Mainly contributes to aromas in wine Many different types of sulphide compounds
Sensorial Terroir	
The smell is not diverse Negative smell: pungent and suffocating	A wide range of aromas Smell can be positive or negative depending on the type of sulphide, concentration and individual preferences

Figure 27: Sulphite versus sulphide

of SO_2 addition can certainly contribute to an irritating odour in wine).

In wine, there are three forms of sulphite depending on the pH: molecular sulphur dioxide (SO_2), bisulphite (HSO_3^-) and sulphite (SO_3^{2-}). In reality, SO_3^{2-} is almost non-existent in the normal range of wine pH, while HSO_3^- is the most abundant form. Sensorially, only the molecular form of SO_2 in wine can contribute to the pungent, irritating or even choking smell as it is the only form that is volatile.

From rotten to luxury

The simplest type of sulphide, as mentioned, is hydrogen sulphide (H_2S). Despite its undecorated chemical structure, an entire book can be written on the significance and functionality of H_2S. In the oxidative world that we inhabit, most living things use oxygen for respiration and for energy production. The reductive sulphur-containing world, on the other hand, is toxic to us. Based on the evidence available, geoscientists suspect that the largest mass extinction on earth 251 million years ago was due to extremely high levels of hydrogen sulphide caused by volcanic eruptions, as well as ancient microbial species that produce mass volumes of sulphides. Therefore, almost all humans and animals are naturally hypersensitive to the smell and even the touch of excessive hydrogen sulphide. A volatile compound like the gaseous form of hydrogen sulphide not only gives a super-pungent smell but also causes irritation to our eyes and skin. It is a sign of danger as well as death (as mentioned in Chapter 8 and Chapter 9). When corpses start to break down further through the action of microorganisms in the absence of oxygen, sulphides like hydrogen sulphide will be generated. This is why rotten egg, which contains and releases a significant amount of hydrogen sulphide, is the best reference and descriptor for the odour produced by this chemical.

In Chapter 9 we saw the general low threshold in the population in terms of detecting the smell of sulphides and the reason should be obvious now. Sulphides, in general, are toxic or even lethal to any organic creatures that live on oxygen to produce energy. The alarm system of the potential danger of sulphides is deeply coded into our DNA. Such sensory properties have significant applications in human society. For example, contrary to our instinct, natural gas that we use for heating or to make fire, with methane as the main compound in it, is equally toxic, yet odourless. In order to warn us of a leak of natural gas, a tiny amount of hydrogen sulphide is mixed in, which we will be able to smell. Therefore, when you detect the rotten egg-like smell of gas, this is from the added 'alarm bell' component, sulphide.

Despite our general revulsion to sulphides in significant concentrations, believe it or not, the right amount of sulphides can be charming. As discussed in Chapter 8, at very low concentrations in our bodies, sulphides like hydrogen sulphide serve as a messenger as well as a regulator of cells. In other words, despite the toxicity of many sulphides, we cannot live without them. Therefore, the smell of sulphides in the safe zone of concentration can be tolerated or even enjoyed. Other than the thiols that show certain fruity and vegetal aromas, there are smells from sulphides that remind us of another type of culinary culture; that of preserved food.

In countries with long culinary histories, preserved food, especially preserved vegetables, have become a national dish. The Korean delicacy kimchi, for example, is now widely appreciated by consumers around the world. The preserved vegetables are prepared in an oxygen-deprived environment so that the microbial activities generate many types of aroma compounds, including assorted sulphides. As a result, kimchi possesses complex smells associated with garlic, green onion, cooked cabbage, etc. These aromas are also found in wines that are considered to have excessive reductive odours, though (coming back to individual perception, preference and expectation again) these odours in wine are judged as faulty.

There are many causes of the obvious reductive, sulphide-related aromas in wine. One cause that is immediately clear is the lack of oxygen influence during winemaking. The other major cause is the type of yeasts involved during fermentation. The majority of sulphides in wine are produced as by-products of yeast activities. Some yeast strains have a strong ability to create those reductive compounds by utilizing compounds like the sulphur-containing amino acids in the grape must. Therefore, it is not difficult to conclude another cause, which is the level of sulphur-containing precursors that yeast can use for producing sulphides. Grape varieties such as Sauvignon Blanc and Syrah tend to accumulate a higher percentage of compounds with elemental sulphur within them (such as glutathione and cysteine). No wonder winemakers always say that Syrah is 'prone to reduction' during winemaking.

Specific sulphide compounds in wine can be cherished in some cases, due to the aroma they provide to certain styles of wine. For example, aroma attributes described as flinty, burnt match-like and somewhat 'mineral' in some ultra-premium white wines can be correlated with sulphide compounds like benzenemethanethiol. Such sensory outcomes tend to be the result of an oxygen-limited style of winemaking. As balancing reduction and oxidation during winemaking involves a great deal of effort and experience, the costs of making those wines are high. On top of this, such 'reductive' wine styles include luxury brands found in regions like Champagne and Burgundy. In these cases, the flint, burnt match-like aromas lend wines a sense of sophistication, leading to praise among wine critics and a premium price for consumers.

Unlike wines, where there are some exceptions as mentioned above, sulphides are enemies to all kinds of spirits because the distillation process can concentrate compounds like hydrogen sulphide to amplify the 'rotten' smells. Therefore, most distillation columns are made from copper, which not only can help heat the entire system to the same temperature, but also because the copper reacts with sulphides to form removable solids – killing two birds with one stone.

Redox and the wine matrix

Redox is a term in chemistry that describes the combined reduction and oxidation reactions. When a reducer is oxidized by the oxidant, the oxidant is also reduced by the reducer. The resulting product is a combination of the former reductive compound and oxidative compound. For example, disulphide with the double sulphur bond (S-S) is a bit more oxidative than sulphide with one sulphur (S) element because the former results from the oxidation of two singular sulphides. Another example is the product of reactions between sulphur dioxide (SO_2) and oxidized phenolics. When phenolics are oxidized to intermediate compounds called quinones, the bisulphites (HSO_3^-) present quickly react with those quinones to inhibit further oxidation. This is the behind-the-scenes, true story of how sulphur dioxide functions as an antioxidant in winemaking. Most of the sensorial attributes of the quinone-bisulphite compound remain to be explored. However, it has been found that the reaction between condensed tannins and bisulphite during wine ageing forms compounds like flavan-3-ol sulphonates, which may result in a less astringent taste on the palate.

One of the key aroma compounds contributing to the typical smell of coffee is 2-furfurylthiol, as discussed in Chapter 11. In wine, this aroma can occur through the combination of the more oxidative furfurals from the oak and the more reductive thiols produced by yeasts. As noted earlier, some Pinotage wines from South Africa smell almost exactly like coffee but without any coffee-related flavouring agents added in. This compound is a typical child of the more reductive 'yin' of thiols and the more oxidative 'yang' of furfurals.

To emphasize one last time: oxidation and reduction happen at the same time. When chemical species A is oxidized by species B, species B is simultaneously reduced by species A. A tiny change in the level of metal ions, alcohol, pH, temperature, pressure and so on can significantly alter the redox potential and the entire composition of oxidative versus reductive compounds, thus making each glass of wine chemically unique.

The power of smelling

Let us conclude by appreciating how powerful the sense of smell is. At the beginning of Part Three, it was pointed out that, anatomically, the emotional centre and the olfactory bulb in our brain are closely connected to one another. Therefore, any aroma compound may have more or less of an impact on our emotions and even our overall health status. This is one of the theories behind aromatherapy, which uses aromatic materials to improve a person's psychological well-being.

Yet, as we have learned in the previous three chapters, it is almost impossible to ensure the outcome of any aromatherapy:

- Different thresholds (Chapter 9) can suggest an individual's complete immunity to certain aroma compounds.
- Odour adaptation (Chapter 10) may quickly reduce our sensitivity towards a target compound over the course of aromatherapy.
- Specific anosmia (Chapter 11) may lead to different responses,

in that an aroma compound might smell nice to one person but trigger a negative response in another.

The emotive and deceptive nature of our sense of smell provides so many joys and sorrows in life, whether or not we are conscious about what compounds were responsible for those experiences.

This is why a wine professional or a food connoisseur may seem to have a more sensitive nose and palate. They are not necessarily (and usually are not) hypertasters as discussed in Chapter 6, but they certainly pay more attention to what kind of tastes and aromas they experience during consumption. Unlocking the power of smell demands the time and effort spent learning a new set of skills called the aroma or odour language. The key is to explore food and beverages as diversely as possible.

Powerful sensations are also difficult to comprehend. Earlier, in Chapter 8, we made a distinction between orthonasal and retronasal olfaction. In terms of retronasal olfaction, the sense of smell works in conjunction with the sense of taste on our palate. So far, only a few scientific methodologies have been used to scratch the surface of this complex taste and odour matrix. While we anticipate further breakthroughs in science that reveal more about the chemical and sensorial terroir of wine aromas, it is important to understand that each look, taste and smell of a glass of wine is sensorially unique to you at any given moment. The more attention you devote to your own perceptions of food and wine, the more potent your eyes, palate and nose will become, ultimately enhancing your enjoyment during eating and drinking occasions.

Final thoughts

Oxidation and reduction reactions permeate every facet of our environment, shaping natural phenomena and every drop of wine in a glass. Miraculously, many volatile compounds formed by those reactions give a vast variety of different smells and humans possess the power to recognize, to judge and to emotionally respond to those aromas.

The discussion of the chemical and sensorial terroir of wine tasting comes to an end. But as Albert Einstein once said: 'The only source of knowledge is experience.' In the epilogue that follows, we will use real wine tasting examples to put all the knowledge in this book into practice.

Epilogue: putting it all together in six pairs of wines

In this epilogue we will bring to life everything we have discussed in the rest of the book. Below, I have recommended six wine pairings that will demonstrate some of the concepts that we have discussed. By tasting the six pairs of wines, you will gain a greater understanding of the chemical and the sensorial terroir of wines. If it is possible to purchase some or all of these wines, I strongly encourage you to taste the two wines in each pair side by side. Comparative tastings like these are the best way of understanding the chemistry of wine in a glass, as well as your own sensory perceptions.

Pair 1: Prosecco and Champagne

Prosecco is a refreshing, light, simple sparkling wine from the Prosecco region of north-east Italy. The wines are fermented and made bubbly in stainless-steel tanks. In comparison, Champagne is a sparkling wine that is generally much more complex and concentrated. Such high quality comes from long ageing with the presence of lees (dead yeast cells) in the bottle, which is one of the key steps of the traditional method. Only those sparkling wines made by the traditional method within the boundaries of France's Champagne region may be called 'Champagne'.

Chapter 3 explained how bubbles are formed in a liquid. With the same pressure in the bottle, Prosecco tends to have larger-sized bubbles than Champagne, most likely due to the higher density and viscosity created by the autolysis of lees in the latter. The other key reason is that Champagne is usually stored for a much longer period of time than Prosecco before the cork is popped. The longer storage time means a loss of pressure in the bottle, leading to smaller bubbles. But always remember that the way the wine encounters the inner wall of the wine glass can also have a notable effect on the quantity of bubbles formed. The more uneven the surface is, the more nucleation sites there are for generating lots of bubbles.

Chapter 3 also addressed the preconceptions that consumers often have towards sparkling wines. Sparkling drinks are strongly associated with celebrations and lively social scenes, making them popular beverages for many occasions. As a result, the appearance of bubbles is the most important element that drives our desire to consume them. In reality, most consumers do not pay attention to the difference between Prosecco and Champagne. As long as it's fizzy and bubbling in the glass, it's a great wine to savour in the moment.

Chapter 4 introduced the umami taste to our tongue. The perfect way of understanding this taste in wine is to savour Prosecco and Champagne side by side. Because of the autolysis of dead yeast cells during the production of Champagne, those sparkling wines contribute to a distinct savoury or umami taste on our palate, while this taste does not feature in Prosecco.

Chapter 5 illustrated the possible mechanisms of perceiving acidity in food and beverages. Both Prosecco and Champagne taste refreshing, but due to the cooler climate, Champagne in general has higher acidity from a chemical point of view. However, the actual sensation of sourness and how long this sensation lasts on the palate depend on many other factors that need further research.

Chapter 5 also brought up the complex interactions of tastes on our palate. With a few exceptions, most Prosecco and Champagne have residual sugars, added at the last stage of winemaking. This sugar addition, called the dosage, helps to ease the sharp acidity in most sparkling wines. Even if the dosage goes up to 12 grams per litre, a Prosecco or a Champagne can taste very dry due to the high acidity and bubbles, which mask our sensation of sweetness. Without any dosage, a Prosecco or a Champagne would taste too sour to some people.

Chapter 7 discussed the tactile sensations experienced in the mouth when drinking wine, including the perception of bubbles (often called the mousse). The bubbles in Champagne often taste creamier and smoother than those in Prosecco. One of the reasons may be that in Champagne, the polysaccharides released by yeast autolysis enrich the mouthfeel to cover the bursting sharpness of bubbles. But it is worth noting the combination of factors that lead to the overall perception of mousse, such as a lower pressure in Champagne due to the longer ageing, as mentioned previously.

Chapter 8 introduced esters as a major group of volatile compounds for fruity aromas. In Prosecco, the expression of esters is prominent, as there are not many other non-ester compounds found in Prosecco that add complexity to the aroma matrix. As a result, Prosecco has much more straightforward apple- and pear-like aromas.

Chapter 12 described the sulphide compounds which can be released by yeasts during fermentation and during lees ageing. As a result, Champagne usually exhibits greater aroma complexity compared to Prosecco, with notes such as flint and burnt match from aroma compounds like benzenemethanethiol, in addition to aromas derived from the esters.

Pair 2: Provence rosé and vintage Port

The pink-coloured wines from Provence, France, tend to be very light in colour. They taste dry and fresh, with simple, fruity aromas. In contrast, vintage Port from Portugal is a deep-coloured red wine. It is made by adding high-strength brandy to stop the fermentation process, making the wine sweet, with high alcohol.

As a heavily extracted red wine, it also contains lots of tannins as well as intense aromas.

Chapter 1 explained anthocyanins in depth, as the key group of compounds that are responsible for the colour of rosé and red wines. The rosé wines from Provence contain low concentrations of anthocyanins, so their colour is very pale. Sensorially, our eyes perceive the pink colour when the saturation of colour is low, meaning that under natural light, the rosé wines look pink due to the 'white' lighting around us. This decreases the saturation of the 'red' light reflected by anthocyanins. A vintage Port, on the other hand, contains not only abundant anthocyanins, but also many stable colour pigments formed by the interactions between anthocyanins and other compounds like tannins. As a result, the colour of Port is deep and long-lasting, since the chemical nature of colour compounds in those wines is stable.

Chapter 3 explained how wine tears form after swirling the wine in a glass. Due to the much higher alcohol content in vintage Port, it displays more prominent tears than you will observe by swirling your glass of Provence rosé. The other thing you will observe in vintage Port is the thick sediments that come out during the later stage of pouring. This is because vintage Port is not filtered during production. The sediments of long-aged vintage Port are thicker as wine pigments become large enough to form visible solids.

Chapter 3 also covered our preconceptions. Nowadays, as pink-coloured wines are in vogue and Ports are treated as old-fashioned, it is much easier to sell Provence rosé than vintage Port wines. The pink colour plays an important role in incentivizing consumers' desire to purchase and taste rosé wines. To invoke the image of summer drinking in the warm Mediterranean holiday destination of Provence, the colour of the local rosé wines is pale, giving the visual impression of being light and refreshing. But note that in countries where pink is not a trendy colour, rosé wines are much less popular. This is again affected by preconceptions.

Chapter 4 talked about the sweet perception in relation to sugar in the wine. Port used to be one of the most popular wines in the world, largely due to its sweetness. Even in the early twentieth century, sugar was not a common commodity. But today, sugar is widely available, to the extent that most of us need to make efforts to reduce or limit our sugar consumption. This is why sweet wines are less highly regarded today, despite an intrinsic human desire for sugar: we can get sugar from plenty of other sources.

Chapter 6 discussed genetic diversity, especially when it comes to the taste of alcohol and tannins. Vintage Port, being high in alcohol and tannins, can taste quite different to different people. Sensitive hypertasters will likely find themselves intolerant of the burning sensation from alcohol and the drying, bitter sensation from tannins found in a young vintage Port, even if for most of us there is sugar to cover or balance those tastes and tactile sensations. As for hypotasters, who have far fewer taste buds, those taste perceptions might not exist at all.

Chapter 7 addressed how anthocyanin and tannin molecules become larger through polymerization over time. As a result, Provence rosé with a handful of anthocyanins can lose the light pink colour rather quickly, leading to a short shelf-life. On the other hand, the large pool of phenolic compounds in a vintage Port becomes more condensed and complex during a long period of ageing, which leads to stable colour as well as modified perception of tannins. In general, the tannins of aged vintage Port taste more well-rounded, but more research needs to be conducted to correlate the chemical composition of phenolics to their mouthfeel.

Chapter 12 illustrated how the oxidation process contributes to a high level of aroma complexity in wine. Provençal rosés are meant to be drunk while the wine is young. Their simple aromas will fade quickly during storage. In contrast, vintage Port has concentrated compounds that aid long-term ageing with oxidative influence. Vintage Port from the 1980s can still taste vibrant, but also over time will have developed a bouquet of oxidative aromas like caramel, toffee, coffee and nuttiness.

Pair 3: Muscat and orange wines

Most wines made from the Muscat family of grape varieties have expressive fruity and floral aromas. Although the varieties within the family are related to each other, the grape skins can vary in colour, leading to white, rosé or even red wines made from different Muscat varieties. Orange wines are technically white wines, but the skins of the white grapes are fermented with the juice. The grape variety (or varieties) can be any, including those in the Muscat family. Because of the extraction from the skins, orange wines have perceivable tannins; the orange or amber colour comes from the oxidation of extracted phenolics. Although not legally defined, many orange wines are also so-called natural wines, which are produced with minimal intervention such as zero addition of sulphites. This deliberately hands-off approach leads to unpredictable microbial activities during winemaking. The resulting aromas may be perceived as strange or even unpleasant by consumers.

Chapter 2 stated that the presence of phenolics facilitates oxidation. Therefore, the defining step of making an orange wine is to ferment with the skins of white grapes. Later, the colour of wine can easily be oxidized to amber or brown. If made using grapes that contain only a small quantity of anthocyanins, such as the pink-coloured Muscat variety, the wine will be closer to the orange colour.

In Chapter 3 we looked at the preconceptions around how cloudy a drink should be. For most white wines, a cloudy appearance seems to be unacceptable to most consumers. But for an orange wine that is also labelled as a natural wine style, people welcome the hazy appearance as it shows the wine is unfiltered and reinforces the impression of low-intervention winemaking.

Chapter 6 addressed the fact that phenolic compounds in wine can be perceived as bitter by certain people. As phenolics are extracted during the winemaking of orange wines, hyper-sensitive tasters may not be able to get used to the sensation of these wines on the palate, just as certain people do not consume beer on a regular basis as they cannot tolerate the bitterness from hops.

Chapter 8 emphasized the importance of terpenoids to all living things. Grapes in the Muscat family are great examples of fruits that nurture the marked aroma compounds that belong to terpenoids. When we smell a glass of Muscat wine, many of the terpenoids express fruity and floral aromas that make the wine appealing.

Chapter 8 also focused on the two pathways of smelling: orthonasal olfaction and retronasal olfaction. The majority of food flavours we 'taste' in the mouth are in fact from the volatile compounds perceived (smelled) via the retronasal pathway. Even though fermentation releases the attractive volatile compounds of terpenoids, many terpenoids are still preserved in the non-volatile form, attached to sugar molecules. The fruity and floral aromas can be more pronounced on the palate than the nose, as our saliva makes the terpenoids volatile and they become detectable retronasally.

Chapter 10 covered the aromas that are not commonly welcomed. While a Muscat wine will express purely fruity and floral scents, some natural-style orange wines can exhibit animal, sticking-plaster or preserved cabbage-like odours. The acceptance of those smells depends on the concentration of the associated volatile compounds as well as personal preference.

Chapter 11 discussed how specific anosmia can affect a person's response to aromas. The volatile isovaleric acid, which can be found in certain natural wines as a by-product of the *Brettanomyces* yeasts, should smell sweaty and cheesy. However, some people are not able to detect the odour of isovaleric acid. Therefore, the same wine can smell either fragrant or objectionable, depending on our genetics.

Pair 4: Sauvignon Blanc and Sauternes

The light-coloured, dry, crisp white wines made from the Sauvignon Blanc variety can show pronounced grapefruit and passion fruit-like aromas, along with vegetal notes like green bell pepper. Sauternes is a region in France that produces wines that often include Sauvignon Blanc grapes, yet they taste lusciously sweet, with a rich, oily mouthfeel. This is because the grapes for making the sweet Sauternes wines are infected by a type of fungus called *Botrytis cinerea*, which in its benevolent form (noble rot) can turn the grape berries into raisins. As a result, wines made from *Botrytis*-affected grapes have concentrated sugar, aromas and many other components.

In Chapter 1 we observed how the skin of white grapes can turn red or purple in colour. Sauternes wines are made from white varieties such as Sauvignon Blanc and Semillon. When those grapes are infected by the special fungus *Botrytis cinerea*, the red-coloured anthocyanins start to accumulate in the grape skins. As a result, Sauternes wines can appear somewhat orange and amber in colour if some skin contact is involved during winemaking.

Chapter 2 also explained the deep colour of Sauternes. If you can manage to find a dry, white Bordeaux wine made from Sauvignon Blanc and a Sauternes from the same vintage, you will see that the former looks much lighter in colour. The reason is that the enzymes produced by *Botrytis* have a strong capability of catalysing the browning process in Sauternes wines.

Chapter 2 covered some other colours that may be observed in white wines. A young wine made from Sauvignon Blanc might have some green hues. This is because the skins of Sauvignon Blanc grapes contain green-coloured chlorophylls, especially if the grapes are picked relatively early for making a light, refreshing style of wine.

Chapter 5 emphasized the deceptive effects on the palate that occur due to taste interactions. If you taste a Sauternes right after having a glass of the dry and crisp Sauvignon Blanc, you might think the lusciously sweet Sauternes lacks acidity. On the other hand, if you taste the Sauternes first, the Sauvignon Blanc will taste far too sour. However, Sauternes does have lots of acidity due to the grape concentration process influenced by *Botrytis*. It is the high level of sugar and other components that spoil our tongue and nose with a sweet perception so that the sour sensation is masked. Having said that, many Sauternes wines do not taste 'sticky' and this is because the sweetness is balanced by a backbone of high acidity.

Chapter 7 explored the less well-studied compounds that may serve important roles in the texture of wine on the palate. The high concentration of sugar is certainly responsible for the rich mouthfeel of Sauternes wines, but compounds such as glycerol and higher alcohols from *Botrytis* infection should also add to the full-bodied, oily sensation.

Chapter 8 introduced the sulphides, which contribute to a wide range of aroma profiles. In particular, the sulphide compound of 3-mercaptohexan-1-ol contributes to the signature grapefruit, passion fruit-like aromas in Sauvignon Blanc wines. But some Sauvignon Blanc wines may also have sulphides like 4-mercapto-4-methylpentan-2-one, which reminds certain individuals of the smell of cat urine.

Chapter 9 discussed the methoxypyrazine compounds, which give vegetal-like odours. These compounds are everywhere in the plant kingdom, but only in varieties like Sauvignon Blanc do methoxypyrazines accumulate in the grape skins. With skin contact during winemaking, the wine will show green bell pepper-like aromas which may or may not be accepted by consumers.

Chapter 9 also highlighted the concept of the odour threshold. There is no absolute number for the threshold of an aroma compound. However, some compounds are notoriously powerful at low concentrations. For example, we are extremely sensitive to the methoxypyrazine compound of 3-isobutyl-2-methoxypyrazine (IBMP). In wines like Sauvignon Blanc, a tiny amount of IBMP can express pronounced green bell pepper-like aromas.

Chapter 11 covered the aromas derived from oak influence. Quite a few Sauternes wines see some new oak influence as the wines are aged in barrel. Therefore, Sauternes wines often express vanilla smells from new oak and even crème brûlée-like aromas resulting from the wine and oak interactions. In comparison, most

Sauvignon Blanc wines are made in an unoaked style, emphasizing fruity and vegetal aromas.

Pair 5: fino Sherry and Syrah

Spain's signature fino Sherry (and manzanilla Sherry) starts out as a dry white wine, but is fortified with high-strength brandy to a specific alcohol level around 15% abv. Then a special status of yeast film called the *flor* will form on top of the wines, creating aroma compounds like acetaldehyde which are not normally found in high concentrations in other wines. In contrast, red wines made from the Syrah variety have a completely different profile, with lots of fruity and spicy aromas. Quite often in young Syrah wines, the smells can be odd due to the reductive compounds.

Chapter 1 identified anthocyanins as the key compounds for the colours of red wines. There are abundant anthocyanins in the skins of Syrah grapes, hence the general deep red to purple colour in many Syrah wines from around the world. Syrah wines from wine regions with a hot climate tend to be more purple in colour. This is because the warm growing environment leads to lower acidity and correspondingly higher pH in the grapes. A higher pH means more anthocyanins are in the quinoidal base form, which appears more blue and purple-coloured.

Chapter 2 pointed out the essential elements of oxidative browning, including the presence of phenolic compounds and oxygen. If a Sherry is deliberately made with some extraction of phenolics and exposed to oxygen influence during years of maturation, it will become the oloroso style of Sherry, which is completely oxidized. Fino Sherry, on the other hand, starts out with free-run juice with a minimal level of phenolics and the thick film of *flor* yeasts protects the wine from being oxidized during maturation. As a result, fino Sherry has a pale colour without any touch of browning.

Chapter 5 demonstrated that the perception of acidity is influenced by multiple chemical and sensorial factors. The chemical data tell us that the concentration of acids in fino Sherry is generally low. However, many fino Sherry wines can give us a crisp mouthfeel. This is due to the depletion of wine components which cover the sour sensation from acids. For example, in order to survive in the absence of sugar, the *flor* yeasts consume glycerol, which could otherwise give a richer mouthfeel and decrease the acidic taste.

Chapter 5 commented on the uncommon salty perception in wine. Fino Sherries tend to taste somewhat salty due to the special process they go through during winemaking. It is possible that the *flor* yeasts produce certain compounds that give an almost salty perception. But it is more likely that the strong, humid wind from the ocean transports salt into the wines during the long maturation of these Sherries.

Chapter 7 discussed the different types of tactile sensations from varied components in wine. Both fino Sherry and wines made from the Syrah grape give sensations that go beyond the five basic tastes on the palate. But fino Sherry is more about the tangy sensation from the *flor* yeasts' influence, whereas Syrah can have high tannins which contribute to the astringent mouthfeel.

Chapter 8 explained that quite a lot of fruity and floral aromas can come from different terpenoid compounds. One tradition of Syrah winemaking is to co-ferment Syrah with up to 20 per cent of the white variety of Viognier. With the inclusion of Viognier, which is rich in terpenoids, Syrah wines can express 'elevated fruitiness and perfumed aromas', as described by some professional wine tasters.

Chapter 11 explored the spicy aromas found in wines. Syrah grapes often accumulate higher levels of a terpenoid compound called rotundone, which is the same compound that is largely responsible for the signature peppery aromas in black and white peppers. Rotundone is also a great example of the specific anosmia discussed in the same chapter. Quite a few studies found a notable proportion of people who are significantly less sensitive to the pepper-like smell of rotundone, while other consumers always find Syrah wines peppery on the nose.

Chapter 12 showcased the oxidative world and the reductive world. The aldehyde compounds in fino Sherry, such as acetaldehyde, contribute to the special aroma profile of this style of Sherry. The *flor* yeasts take up alcohols as nutrients and transform them into a more oxidative status in the form of aldehydes. Once the alcohols are oxidized to aldehydes, their aroma thresholds decrease, meaning it is easier for our nose to perceive the more oxidative aldehydes. On the other hand, a young Syrah red wine tends to have sulphide compounds which are more reductive in their chemical nature. As indicated in the chapter, there are thousands of sulphide compounds with diverse aromas. They can be fruity, mineral, meaty or pungent. Don't be surprised if the Syrah wine in front of you smells like rotten eggs. In most cases, those aromas should disappear after the bottle has been opened for a while, because the sulphide compounds will be oxidized and lose their initial aroma properties.

Pair 6: Beaujolais nouveau and Rioja Gran Reserva

Beaujolais is a region in France that is known for its Beaujolais nouveau, simple, fruity red wines made from the Gamay variety and released in the November after harvest. (Note that there are also very high-quality Beaujolais wines with more tannins, body and aroma complexity from the grapes as well as from the stems.) The Rioja region in Spain can also produce easy-drinking wines, but the Gran Reserva red wines from this region have always been through at least two years of maturation in oak barrels and at least three years of ageing in bottle, which modifies the colour, changes the tannin structure and incubates lots of matured aromas.

Chapter 1 described the ever-changing chemical nature of anthocyanins. As red wine ages, anthocyanins and other phenolic compounds can polymerize and become condensed pigments. Beaujolais nouveau mainly consists of singular anthocyanin molecules due to its youth. Those anthocyanins are susceptible to sulphite bleaching and changes in pH. At low pH, the wine appears to be red in colour, whereas at high pH, the wine can be more purple in colour. In contrast, a Rioja Gran Reserva red wine contains stable colour pigments in the form of condensed anthocyanin and tannin polymers. The much older Rioja Gran Reserva may also appear to have a garnet, orange or tawny hue. One of the elements that contributes to such colour can be the xanthylium pigment, which develops through long-term ageing and it appears more yellow in colour.

Chapter 3 discussed sediments in long-aged red wines. In old Rioja Gran Reserva wines, the colour pigments have become large enough to form sediments rather than being dissolved in the liquid. Consequently, the colour of the wines may be lighter, but the colour compounds are more stable compared to those in young Beaujolais wines.

Chapter 7 proposed the infinite possibilities of the chemical structure of tannins. Therefore, we could only reference some chemical data and trust our own palate when perceiving tannins in wine. Beaujolais nouveau is made by a technique called carbonic maceration, which involves a quick extraction process that results in a low concentration of tannins. Rioja Gran Reserva, on the other hand, starts out with the deep extraction of tannins to have a sufficient phenolic pool for long-term ageing. After years of maturation and storage, the tannins are polymerized to become long-chain, large tannin molecules. When tasting the two wines in this pair, it should be obvious that the Beaujolais nouveau has a smooth sensation with little astringent mouthfeel from tannins. The tannins of the Rioja Gran Reserva can be round and soft due to ageing, but a high concentration of the large tannin molecules fulfils the entire palate, leaving a tactile sensation that is structured and powerful. Note that if you choose to age these wines for decades, the tannin perception may become weak, as most of the anthocyanins and tannins will precipitate in the wine bottle to become sediment.

Chapter 8 illustrated the origin, formation and application of esters as key aroma compounds. Beaujolais nouveau wines are among the best red wines to demonstrate what esters can smell like. The carbonic maceration technique used in making such wines creates esters as the predominant aroma compounds. If you perceive banana, candy-like aromas in Beaujolais nouveau, you are in the right ballpark, as bananas are full of esters and most candies are flavoured using ester compounds.

Chapter 9 covered the materials other than the grapes themselves that can contribute to obvious aromas in wine. Other than the commonly used oak materials, stems and/or leaves can be (legally) included during winemaking, whether it's done intentionally or not. As mentioned, the higher quality Beaujolais wines, especially the single *cru* wines from Beaujolais, can include stems during fermentation. The stems can add another layer of aromas such as fresh herbs, dried herbs and seeds as cooking spices.

Chapter 11 mapped out the aroma compounds of oak. Compared to most Beaujolais wines, Rioja Gran Reserva receives significantly more new oak influence, which adds aroma complexity. Newly-made oak barrels contain many volatile compounds such as vanilla-like vanillin, smoke-like guaiacol and coconut-like oak lactones. Traditionally, Rioja Gran Reserva has used American oak, which has a significantly higher level of oak lactones compared to French oak. Therefore, the traditional style of Rioja Gran Reserva tends to express more coconut-like aromas that resemble certain smells in bourbon whiskeys.

Chapter 12 explained the mechanisms of oxidation in wine. The long ageing of Rioja Gran Reserva develops the sensorial outcome of oxidative ageing. Other than the change of colour and tannins as previously discussed, the aromas are also transformed. There are more raisin, caramel and nutty aromas, which cannot be found in youthful wines like Beaujolais nouveau. The magic of wine ageing remains to be explored by the world of science.

A final note

Please do not get me wrong. When I taste a glass of wine, I do not force myself to think about the chemical components in wine, nor do I consider how my five senses function during tasting. I simply enjoy drinking what is in front of me. First and foremost, I rarely drink a glass of wine by myself. To me, the majority of enjoyment of wine is about sharing it with others, whether they are family, friends or strangers.

It is when we enjoy wines together that the miracles of terroir come to life, prompting me to think about everything behind a glass and discuss it among good company. As mentioned in the Introduction, the modern concept of terroir spans a wide array of elements in the world of wine: history, land ownership, contracts, grape growing, winemaking, marketing, trading, transportation, legislation, psychology, economy, ecology, culture and more. This book delves specifically into the realms of chemical and sensorial terroir, subjects that have prominently featured in my discussions during tastings – likely because of my professional involvement in these fields. Again, these topics did not initially occupy my thoughts but instead emerged from conversations with others during both informal and formal tasting sessions.

While my primary writing focus is the science behind wine tasting, particularly addressing the chemical and sensorial aspects of terroir, I remain actively involved in reading about all other facets of wine. This pursuit aims to enrich the quality of conversations during tastings throughout my life. In essence, I hope that this book, along with the wealth of other wine books available, will add more colours, tastes and aromas to your tasting experiences of all foods and beverages.

References

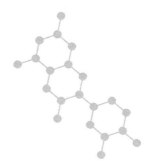

Chapter 1

Casassa, L. F., Keirsey, L. S., Mireles, M. S., & Harbertson, J. F. (2012). 'Cofermentation of Syrah with Viognier: Evolution of Color and Phenolics during Winemaking and Bottle Aging'. *American Journal of Enology and Viticulture*, 63(4), 538–43.

de Freitas, V., & Mateus, N. (2011). 'Formation of Pyranoanthocyanins in Red Wines: A New and Diverse Class of Anthocyanin Derivatives'. *Analytical and Bioanalytical Chemistry*, 401, 1463–73.

Fu, X.-Z., Zhang, Q.-A., Zhang, B.-S., & Liu, P. (2018). 'Effect of Ultrasound on the Production of Xanthylium Cation Pigments in a Model Wine'. *Food Chemistry*, 268, 431–40.

Hecht, S. (1937). 'Rods, Cones and the Chemical Basis of Vision'. *Physiological Reviews*, 17(2), 239–90.

Ibraheem, N. A., Hasan, M., Khan, R. & Mishra, P. K. (2012). 'Understanding Color Models: A Review'. *ARPN Journal of Science and Technology*, 2, 265–75.

Lapidot, T., Harel, S., Akiri, B., Granit, R., & Kanner, J. (1999). 'pH-Dependent Forms of Red Wine Anthocyanins as Antioxidants'. *Journal of Agricultural and Food Chemistry*, 47(1), 67–70.

Nassau, K. (2001). *The Physics and Chemistry of Color: The Fifteen Causes of Color* (2nd ed.). Wiley.

Vivar-Quintana, A. M., Santos-Buelga, C., & Rivas-Gonzalo, J. C. (2002). 'Anthocyanin-derived Pigments and Colour of Red Wines'. *Analytica Chimica Acta*, 458(1), 147–55.

Waterhouse, A. L., Sacks, G. L., & Jeffery, D. W. (2016). *Understanding Wine Chemistry*. Wiley.

Waterhouse, A. L., & Zhu, J. (2019). 'A Quarter Century of Wine Pigment Discovery'. *Journal of the Science of Food and Agriculture*, 100(14), 5093–101.

Chapter 2

Deeb, S. (2005). 'The Molecular Basis of Variation in Human Color Vision'. *Clinical Genetics*, 67, 369–77.

Fernandez-Zurbano, P., Ferreira, V., Escudero, A., & Cacho, J. (1998). 'Role of Hydroxycinnamic Acids and Flavanols in the Oxidation and Browning of White Wines'. *Journal of Agricultural and Food Chemistry*, 46(12), 4937–44.

Fernandez-Zurbano, P., Ferreira, V., Pena, C., et al. (1995). 'Prediction of Oxidative Browning in White Wines as a Function of Their Chemical Composition'. *Journal of Agricultural and Food Chemistry*, 43(11), 2813–17.

Khoo, H.-E., Prasad, K. N., Kong, K.-W., Jiang, Y., & Ismail, A. (2011). 'Carotenoids and Their Isomers: Color Pigments in Fruits and Vegetables'. *Molecules*, 16, 1710–38.

Kim, H. J., Ryou, J. H., Choi, K. T., et al. (2022). 'Deficits in Color Detection in Patients with Alzheimer Disease'. *PLoS ONE*, 17(1), e0262226.

Martinez, M. V., & Whitaker, J. R. (1995). 'The Biochemistry and Control of Enzymatic Browning'. *Trends in Food Science & Technology*, 6(6), 195–200.

Merbs, S., & Nathans, J. (1992). 'Absorption Spectra of Human Cone Pigments'. *Nature*, 356, 433–5.

Milne, B. F., Toker, Y., Rubio, A., & Nielsen, S. B. (2015). 'Unraveling the Intrinsic Color of Chlorophyll'. *Angewandte Chemie International Edition*, 54, 2170–3.

Neitz, J., & Neitz, M. (2011). 'The Genetics of Normal and Defective Color Vision'. *Vision Research*, 51(7), 633–51.

Oliveira, C. M., Ferreira, A. C. S., De Freitas, V., & Silva, A. M. S. (2011). 'Oxidation Mechanisms Occurring in Wines'. *Food*

Research International, 44(5), 1115–26.

Oszmianski, J., Cheynier, V., & Moutounet, M. (1996). 'Iron-catalyzed Oxidation of (+)-Catechin in Model Systems'. *Journal of Agricultural and Food Chemistry*, 44(7), 1712–15.

Roy, M. S., Podgor, M. J., Collier, B., et al. (1991). 'Color Vision and Age in a Normal North American Population'. *Graefe's Archive for Clinical and Experimental Ophthalmology*, 229, 139–44.

Sanocki, E., Shevell, S. K., & Winderickx, J. (1994). 'Serine/alanine Amino Acid Polymorphism of the L-cone Photopigment Assessed by Dual Rayleigh-type Color Matches'. *Vision Research*, 34(3), 377–82.

Singleton, V. L., & Kramling, T. E. (1976). 'Browning of White Wines and an Accelerated Test for Browning Capacity'. *American Journal of Enology and Viticulture*, 27(4), 157–60.

Wolff, B. E., Bearse, M. A., Schneck, M. E., et al. (2015). 'Color Vision and Neuroretinal Function in Diabetes'. *Documenta Ophthalmologica*, 130, 131–9.

Chapter 3

Blackmore, H., Hidrio, C., & Yeomans, M. R. (2022). 'How Sensory and Hedonic Expectations Shape Perceived Properties of Regular and Non-alcoholic Beer'. *Food Quality and Preference*, 99, 104562.

Donadini, G., Fumi, M. D., & de Faveri, M. D. (2011). 'How Foam Appearance Influences the Italian Consumer's Beer Perception and Preference'. *Journal of the Institute of Brewing*, 117, 523–33.

Duncker, K. (1939). 'The Influence of Past Experience upon Perceptual Properties'. *The American Journal of Psychology*, 52(2), 255–65.

Garber, L. L. Jr., Hyatt, E. M., & Starr, R. G. Jr. (2000). 'The Effects of Food Color on Perceived Flavor'. *Journal of Marketing Theory and Practice*, 8(4), 59–72.

Jaros, D., Thamke, I., Raddatz, H., et al. (2009). 'Single-cultivar

Cloudy Juice Made from Table Apples: An Attempt to Identify the Driving Force for Sensory Preference'. *European Food Research and Technology*, 229, 51–61.

King, E. S., Dunn, R. L., & Heymann, H. (2013). 'The Influence of Alcohol on the Sensory Perception of Red Wines'. *Food Quality and Preference*, 28(1), 235–43.

Koch, C., & Koch, E. C. (2003). 'Preconceptions of Taste Based on Color'. *The Journal of Psychology*, 137(3), 233–42.

Lick, E., König, B., Kpossa, M. R., & Buller, V. (2017). 'Sensory Expectations Generated by Colors of Red Wine Labels'. *Journal of Retailing and Consumer Services*, 37, 146–58.

Morrot, G., Brochet, F., & Dubourdieu, D. (2001). 'The Color of Odors'. *Brain and Language*, 79, 309–20.

Okamoto, M., & Dan, I. (2013). 'Extrinsic Information Influences Taste and Flavor Perception: A Review from Psychological and Neuroimaging Perspectives'. *Seminars in Cell & Developmental Biology*, 24(3), 247–55.

Reinoso-Carvalho, F., Dakduk, S., Wagemans, J., & Spence, C. (2019). 'Dark vs. Light Drinks: The Influence of Visual Appearance on the Consumer's Experience of Beer'. *Food Quality and Preference*, 74, 21–9.

Shankar, M. U., Levitan, C. A., Prescott, J., et al. (2009). 'The Influence of Color and Label Information on Flavor Perception'. *Chemical Senses*, 2, 53–8.

Siegrist, M., & Cousin, M.-E. (2009). 'Expectations Influence Sensory Experience in a Wine Tasting'. *Appetite*, 52(3), 762–5.

Smith, A. M., & McSweeney, M. B. (2019). 'Partial Projective Mapping and Ultra-flash Profile with and without Red Light: A Case Study with White Wine'. *Journal of Sensory Studies*, 34, e12528.

Spence, C. (2018). 'What is so Unappealing About Blue Food and Drink?'. *International Journal of Gastronomy and Food Science*, 14, 1–8.

Veale, R., & Quester, P. (2008). 'Consumer Sensory Evaluations of Wine Quality: The Respective Influence of Price and Country of Origin'. *Journal of Wine Economics*, 3(1), 10–29.

Włodarska, K., Pawlak-Lemańska, K., Górecki, T., & Sikorska, E. (2016). 'Perception of Apple Juice: A Comparison of Physicochemical Measurements, Descriptive Analysis and Consumer Responses'. *Journal of Food Quality*, 39, 351–61.

Zampini, M., & Spence, C. (2005). 'Modifying the Multisensory Perception of a Carbonated Beverage Using Auditory Cues'. *Food Quality and Preference*, 16(7), 632–41.

Chapter 4

Bachmanov, A. A., Bosak, N. P., Lin, C., et al. (2014). 'Genetics of Taste Receptors'. *Chemical Senses*, 20(16), 2669–83.

Boring, E. G. (1942). *Sensation and Perception in the History of Experimental Psychology*. Appleton-Century.

Gawel, R., Smith, P. A., & Waters, E. J. (2016). 'Influence of Polysaccharides on the Taste and Mouthfeel of White Wine'. *Australian Journal of Grape and Wine Research*, 22, 350–7.

Haagen-Smit, A. J. (1952). 'Smell and Taste'. *Scientific American*, 186(3), 28–32.

Hanig, D. P. (1901). 'Zur Psychophysik des Geschmackssinnes'. *Philosophische Studien*, 17, 576–623.

Klosse, P. (2013). 'Umami in Wine'. *Research in Hospitality Management*, 2(1–2), 25–8.

Noble, A. C., & Bursick, G. F. (1984). 'The Contribution of Glycerol to Perceived Viscosity and Sweetness in White Wine'. *American Journal of Enology and Viticulture*, 35, 110–12.

Schmidt, C. V., Olsen, K., & Mouritsen, O. G. (2021). 'Umami Potential of Fermented Beverages: Sake, Wine, Champagne and Beer'. *Food Chemistry*, 360, 128971.

Spence, C. (2022). 'The Tongue Map and the Spatial Modulation of Taste Perception'. *Current Research in Food Science*, 5, 598–610.

Wang, W., Zhou, X., & Liu, Y. (2020). 'Characterization and Evaluation of Umami Taste: A Review'. *TrAC Trends in Analytical Chemistry*, 127, 115876.

Chapter 5

Ahne, G., Erras, A., Hummel, T., & Kobal, G. (2000). 'Assessment of Gustatory Function by Means of Tasting Tablets'. *The Laryngoscope*, 110, 1396–1401.

Amerine, M. A., Roessler, E. B., & Ough, C. S. (1965). 'Acids and the Acid Taste. I. The Effect of pH and Titratable Acidity'. *American Journal of Enology and Viticulture*, 16(1), 29–37.

Boulton, R. (1980). 'The Relationships Between Total Acidity, Titratable Acidity and pH in Wine'. *American Journal of Enology and Viticulture*, 31(1), 76–80.

Breslin, P. A., & Beauchamp, G. K. (1995). 'Suppression of Bitterness by Sodium: Variation Among Bitter Taste Stimuli'. *Chemical Senses*, 20, 609–23.

Forestell, C. A., & Mennella, J. A. (2017). 'The Relationship Between Infant Facial Expressions and Food Acceptance'. *Current Nutrition Reports*, 6, 141–7.

Heymann, H., Hopfer, H., & Bershaw, D. (2014). 'Minerality in White Wines'. *Journal of Sensory Studies*, 29, 1–13.

Keast, R. S. J., & Breslin, P. A. S. (2003). 'An Overview of Binary Taste-taste Interactions'. *Food Quality and Preference*, 14(2), 111–24.

Morata, A., Loira, I., Tesfaye, W., et al. (2018). 'Lachancea Thermotolerans Applications in Wine Technology'. *Fermentation*, 4(3), 53.

Murata, Y., Kataoka-Shirasugi, N., & Amakawa, T. (2002). 'Electrophysiological Studies of Salty Taste Modification by Organic Acids in the Labellar Taste Cell of the Blowfly'. *Chemical Senses*, 27(1), 57–65.

O'Mahony, M., Goldenberg, M., Stedmon, J., & Alford, J. (1979). 'Confusion in the Use of the Taste Adjectives "Sour" and "Bitter"'. *Chemical Senses*, 4(4), 301–18.

Parr, W. V., Ballester, J., Peyron, D., Grose, C., & Valentin, D. (2015). 'Perceived Minerality in Sauvignon Wines: Influence of Culture and Perception Mode'. *Food Quality and Preference*, 41, 121–32.

Puri, S., & Lee, Y. (2021). 'Salt Sensation and Regulation'. *Metabolites*, 11(3), 175.

Purves, D., Augustine, G. J., Fitzpatrick, D., et al. (Eds) (2001). *Neuroscience* (2nd ed.). Sinauer Associates.

Sadler, G. D., & Murphy, P. A. (2010). 'pH and Titratable Acidity'. *Food Analysis*, 4, 219–38.

Theunissen, M. J. M., Polet, I. A., Kroeze, J. H. A., & Schifferstein, H. N. J. (2000). 'Taste Adaptation During the Eating of Sweetened Yogurt'. *Appetite*, 34(1), 21–7.

Chapter 6

Arnold, R. A., Noble, A. C., & Singleton, V. L. (1980). 'Bitterness and Astringency of Phenolic Fractions in Wine'. *Journal of Agricultural and Food Chemistry*, 28(3), 675–8.

Behrens, M., & Meyerhof, W. (2013). 'Bitter Taste Receptor Research Comes of Age: From Characterization to Modulation of TAS2Rs'. *Seminars in Cell & Developmental Biology*, 24(3), 215–21.

Matsuo, R. (2000). 'Role of Saliva in the Maintenance of Taste Sensitivity'. *Critical Reviews in Oral Biology & Medicine*, 11(2), 216–29.

Mojet, J., Christ-Hazelhof, E., & Heidema, J. (2001). 'Taste Perception with Age: Generic or Specific Losses in Threshold Sensitivity to the Five Basic Tastes?'. *Chemical Senses*, 26(7), 845–60.

Noble, A. C. (1994). 'Bitterness in Wine'. *Physiology & Behavior*, 56(6), 1251–5.

Nolden, A. A., & Hayes, J. E. (2015). 'Perceptual Qualities of Ethanol Depend on Concentration and Variation in These Percepts Associates with Drinking Frequency'. *Chemical Senses*, 8, 149–57.

Pickering, G. J., Simunkova, K., & DiBattista, D. (2004). 'Intensity of Taste and Astringency Sensations Elicited by Red Wines is Associated with Sensitivity to PROP (6-n-propylthiouracil)'. *Food Quality and Preference*, 15(2), 147–54.

Shi, P., Zhang, J., Yang, H., & Zhang, Y. P. (2003). 'Adaptive Diversification of Bitter Taste Receptor Genes in Mammalian Evolution'. *Molecular Biology and Evolution*, 20(5), 805–14.

Yang, Q., Williamson, A.-M., Hasted, A., & Hort, J. (2020). 'Exploring the Relationships Between Taste Phenotypes, Genotypes, Ethnicity, Gender and Taste Perception Using Chi-square and Regression Tree Analysis'. *Food Quality and Preference*, 83, 103928.

Yoshinaka, M., Ikebe, K., Uota, M., et al. (2016). 'Age and Sex Differences in the Taste Sensitivity of Young Adult, Young-old and Old-old Japanese'. *Geriatrics & Gerontology International*, 16, 1281–8.

Chapter 7

Cheynier, V., Duenas-Paton, M., Salas, E., et al. (2006). 'Structure and Properties of Wine Pigments and Tannins'. *American Journal of Enology and Viticulture*, 57(3), 298–305.

Gawel, R., & Waters, E. J. (2008). 'The Effect of Glycerol on the Perceived Viscosity of Dry White Table Wine'. *Journal of Wine Research*, 19(2), 109–14.

Harbertson, J. F., Parpinello, G. P., Heymann, H., & Downey, M. O. (2012). 'Impact of Exogenous Tannin Additions on Wine Chemistry and Wine Sensory Character'. *Food Chemistry*, 131(3), 999–1008.

Ishikawa, T., & Noble, A. C. (1995). 'Temporal Perception of Astringency and Sweetness in Red Wine'. *Food Quality and Preference*, 6(1), 27–33.

Liem, D. G. (2017). 'Infants' and Children's Salt Taste Perception and Liking: A Review'. *Nutrients*, 9, 1011.

Mennella, J. A., Bobowski, N. K., & Reed, D. R. (2016). 'The Development of Sweet Taste: From Biology to Hedonics'. *Reviews in Endocrine and Metabolic Disorders*, 17, 171–8.

Noble, A. C., & Bursick, G. F. (1984). 'The Contribution of Glycerol to Perceived Viscosity and Sweetness in White Wine'. *American Journal of Enology and Viticulture*, 35, 110–12.

Pickering, G. J., Heatherbell, D. A., Vanhanen, L. P., & Barnes, M. F. (1998). 'The Effect of Ethanol Concentration on the Temporal Perception of Viscosity and Density in White Wine'. *American Journal of Enology and Viticulture*, 49, 306–18.

Rankine, B. C., & Bridson, D. A. (1971). 'Glycerol in Australian Wines and Factors Influencing its Formation'. *American Journal of Enology and Viticulture*, 22(1), 6–12.

Ross, C. F., & Weller, K. (2008). 'Effect of Serving Temperature on the Sensory Attributes of Red and White Wines'. *Journal of Sensory Studies*, 23, 398–416.

Singleton, V. L. (1992). 'Tannins and the Qualities of Wines'. In *Plant Polyphenols: Synthesis, Properties, Significance* (pp. 859–80). Springer US.

Chapter 8

Black, C. A., Parker, M., Siebert, T. E., Capone, D. L., & Francis, I. L. (2015). 'Terpenoids: Role in Wine Flavour'. *Australian Journal of Grape and Wine Research*, 21, 582–600.

Comuzzo, P., Tat, L., Tonizzo, A., & Battistutta, F. (2006). 'Yeast Derivatives (Extracts and Autolysates) in Winemaking: Release of Volatile Compounds and Effects on Wine Aroma Volatility'. *Food Chemistry*, 99(2), 217–30.

Kamatou, G. P. P., & Viljoen, A. M. (2008). 'Linalool – a Review of a Biologically Active Compound of Commercial Importance'. *Natural Product Communications*, 3(7), 1183–92.

Oliveira, C., Barbosa, A., Ferreira, A. C. S., Guerra, J., & Guedes DE Pinho, P. (2006). 'Carotenoid Profile in Grapes Related to Aromatic Compounds in Wines from Douro Region'. *Journal of Food Science*, 71, S1–S7.

Ong, P. K. C., & Acree, T. E. (1999). 'Similarities in the Aroma Chemistry of Gewürztraminer Variety Wines and Lychee (*Litchi chinesis Sonn.*) Fruit'. *Journal of Agricultural and Food Chemistry*, 47(2), 665–70.

Pineau, B., Barbe, J.-C., Van Leeuwen, C., & Dubourdieu, D. (2007). 'Which Impact for β-Damascenone on Red Wines

Aroma?' *Journal of Agricultural and Food Chemistry*, 55(10), 4103–8.

Rozin, P. (1982). '"Taste-smell Confusions" and the Duality of the Olfactory Sense'. *Perception & Psychophysics*, 31, 397–401.

Small, D. M., Gerber, J. C., Mak, Y. E., & Hummel, T. (2005). 'Differential Neural Responses Evoked by Orthonasal Versus Retronasal Odorant Perception in Humans'. *Neuron*, 47(4), 593–605.

Tomasino, E., & Bolman, S. (2021). 'The Potential Effect of β-Ionone and β-Damascenone on Sensory Perception of Pinot Noir Wine Aroma'. *Molecules*, 26, 1288.

Villamor, R. R., & Ross, C. F. (2013). 'Wine Matrix Compounds Affect Perception of Wine Aromas'. *Annual Review of Food Science and Technology*, 4(1), 1–20.

Williams, P. J., Strauss, C. R., & Wilson, B. (1981). 'Classification of the Monoterpenoid Composition of Muscat Grapes'. *American Journal of Enology and Viticulture*, 32, 230–5.

Chapter 9

Benkwitz, F., Nicolau, L., Lund, C., et al. (2012). 'Evaluation of Key Odorants in Sauvignon Blanc Wines Using Three Different Methodologies'. *Journal of Agricultural and Food Chemistry*, 60(25), 6293–302.

Guth, H. (1997). 'Quantitation and Sensory Studies of Character Impact Odorants of Different White Wine Varieties'. *Journal of Agricultural and Food Chemistry*, 45(8), 3027–32.

Heymann, H., Noble, A. C., & Boulton, R. B. (1986). 'Analysis of Methoxypyrazines in Wines. 1. Development of a Quantitative Procedure'. *Journal of Agricultural and Food Chemistry*, 34(2), 268–71.

Lawless, H. T., & Heymann, H. (2010). *Sensory Evaluation of Food: Principles and Practices*. Springer Science & Business Media.

Lund, C. M., Thompson, M. K., Benkwitz, F., et al. (2009). 'New Zealand Sauvignon Blanc Distinct Flavor Characteristics:

Sensory, Chemical and Consumer Aspects'. *American Journal of Enology and Viticulture*, 60(1), 1–12.

Meillon, S., Urbano, C., & Schlich, P. (2009). 'Contribution of the Temporal Dominance of Sensations (TDS) Method to the Sensory Description of Subtle Differences in Partially Dealcoholized Red Wines'. *Food Quality and Preference*, 20(7), 490–9.

Preston, L. D., Block, D. E., Heymann, H., et al. (2008). 'Defining Vegetal Aromas in Cabernet Sauvignon Using Sensory and Chemical Evaluations'. *American Journal of Enology and Viticulture*, 59, 137–45.

Ruth, J. H. (1986). 'Odor Thresholds and Irritation Levels of Several Chemical Substances: A Review'. *American Industrial Hygiene Association Journal*, 47(3), A-142–A-151.

Shibamoto, T. (1986). 'Odor Threshold of Some Pyrazines'. *Journal of Food Science*, 51, 1098–9.

Tempere, S., Cuzange, E., Malak, J., et al. (2011). 'The Training Level of Experts Influences Their Detection Thresholds for Key Wine Compounds'. *Chemical Senses*, 4, 99–115.

Zhang, L., Tao, Y. S., Wen, Y., & Wang, H. (2013). 'Aroma Evaluation of Young Chinese Merlot Wines with Denomination of Origin'. *South African Journal of Enology and Viticulture*, 34(1), 46–53.

Chapter 10

Buser, H. R., Zanier, C., & Tanner, H. (1982). 'Identification of 2,4,6-Trichloroanisole as a Potent Compound Causing Cork Taint in Wine'. *Journal of Agricultural and Food Chemistry*, 30(2), 359–62.

Dalton, P., & Wysocki, C. J. (1996). 'The Nature and Duration of Adaptation Following Long-term Odor Exposure'. *Perception & Psychophysics*, 58(5), 781–92.

Fu, S. G., Yoon, Y., & Bazemore, R. (2002). 'Aroma-active Components in Fermented Bamboo Shoots'. *Journal of Agricultural and Food Chemistry*, 50(3), 549–54.

Grosofsky, A., Haupert, M. L., & Versteeg, S. W. (2011). 'An Exploratory Investigation of Coffee and Lemon Scents and Odor Identification'. *Perceptual and Motor Skills*, 112(2), 536–8.

Heresztyn, T. (1986). 'Formation of Substituted Tetrahydropyridines by Species of Brettanomyces and Lactobacillus Isolated from Mousy Wines'. *American Journal of Enology and Viticulture*, 37(2), 127–32.

Jeong, H. S., & Ko, Y. T. (2010). 'Major Odor Components of Raw Kimchi Materials and Changes in Odor Components and Sensory Properties of Kimchi During Ripening'. *Journal of the Korean Society of Food Culture*, 25(5), 607–14.

Macku, C., Gonzalez, L., Schleussner, C., et al. (2009). 'Sensory Screening for Large-Format Natural Corks by "Dry Soak" Testing and Its Correlation to Headspace Solid-Phase Microextraction (SPME) Gas Chromatography/Mass Spectrometry (GC/MS) Releasable Trichloroanisole (TCA) Analysis'. *Journal of Agricultural and Food Chemistry*, 57(17), 7962–8.

Romano, A., Perello, M. C., Revel, G. D., & Lonvaud-Funel, A. (2008). 'Growth and Volatile Compound Production by Brettanomyces/Dekkera Bruxellensis in Red Wine'. *Journal of Applied Microbiology*, 104(6), 1577–85.

Romano, A., Perello, M. C., Lonvaud-Funel, A., Sicard, G., & de Revel, G. (2009). 'Sensory and Analytical Re-evaluation of "Brett character"'. *Food Chemistry*, 114(1), 15–19.

Snowdon, E. M., Bowyer, M. C., Grbin, P. R., & Bowyer, P. K. (2006). 'Mousy Off-flavor: A Review'. *Journal of Agricultural and Food Chemistry*, 54(18), 6465–74.

Wang, J., & Luca, V. D. (2005). 'The Biosynthesis and Regulation of Biosynthesis of Concord Grape Fruit Esters, Including 'Foxy' Methylanthranilate'. *The Plant Journal*, 44(4), 606–19.

Wedral, D., Shewfelt, R., & Frank, J. (2010). 'The Challenge of Brettanomyces in Wine'. *LWT-Food Science and Technology*, 43(10),1474–9.

Chapter 11

Amoore, J. E. (1967). 'Specific Anosmia: A Clue to the Olfactory Code'. *Nature*, 214(5093), 1095–8.

Croy, I., Olgun, S., Mueller, L., et al. (2016, May). 'Spezifische Anosmie als Prinzip olfaktorischer Wahrnehmung [Specific Anosmia as a Principle of Olfactory Perception]'. *HNO*, 64(5), 292–5.

Cutzach, I., Chatonnet, P., Henry, R., & Dubourdieu, D. (1997). 'Identification of Volatile Compounds with a "Toasty" Aroma in Heated Oak Used in Barrelmaking'. *Journal of Agricultural and Food Chemistry*, 45(6), 2217–24.

Dorries, K. M., Schmidt, H. J., Beauchamp, G. K., & Wysocki, C. J. (1989). 'Changes in Sensitivity to the Odor of Androstenone During Adolescence'. *Developmental Psychobiology: The Journal of the International Society for Developmental Psychobiology*, 22(5), 423–35.

Francis, I. L., Sefton, M. A., & Williams, P. J. (1992). 'A Study by Sensory Descriptive Analysis of the Effects of Oak Origin, Seasoning and Heating on the Aromas of Oak Model Wine Extracts'. *American Journal of Enology and Viticulture*, 43(1), 23–30.

Gaby, J. M., Bakke, A. J., Baker, A. N., Hopfer, H., & Hayes, J. E. (2020). 'Individual Differences in Thresholds and Consumer Preferences for Rotundone Added to Red Wine'. *Nutrients*, 12(9), 2522.

Jarauta, I., Cacho, J., & Ferreira, V. (2005). 'Concurrent Phenomena Contributing to the Formation of the Aroma of Wine During Aging in Oak Wood: An Analytical Study'. *Journal of Agricultural and Food Chemistry*, 53(10), 4166–77.

Keller, A., Zhuang, H., Chi, Q., Vosshall, L. B., & Matsunami, H. (2007). 'Genetic Variation in a Human Odorant Receptor Alters Odour Perception'. *Nature*, 449(7161), 468–72.

Menashe, I., Man, O., Lancet, D., & Gilad, Y. (2003). 'Different Noses for Different People'. *Nature Genetics*, 34(2), 143–4.

Menashe, I., Abaffy, T., Hasin, Y., et al. (2007). 'Genetic

Elucidation of Human Hyperosmia to Isovaleric Acid'. *PLoS Biology*, 5(11), e284.

Plotto, A., Barnes, K. W., & Goodner, K. L. (2006). 'Specific Anosmia Observed for β-ionone, But not for α-ionone: Significance for Flavor Research'. *Journal of Food Science*, 71(5), S401–S406.

Serby, M., Larson, P., & Kalkstein, D. (1991). 'The Nature and Course of Olfactory Deficits in Alzheimer's Disease'. *The American Journal of Psychiatry*, 148(3), 357–60.

Siebert, T. E., Wood, C., Elsey, G. M., & Pollnitz, A. P. (2008). 'Determination of Rotundone, the Pepper Aroma Impact Compound, in Grapes and Wine'. *Journal of Agricultural and Food Chemistry*, 56(10), 3745–8.

Tempere, S., Cuzange, E., Malak, J., et al. (2011). 'The Training Level of Experts Influences Their Detection Thresholds for Key Wine Compounds'. *Chemosensory Perception*, 4, 99–115.

Wood, C., Siebert, T. E., Parker, M., et al. (2008). 'From Wine to Pepper: Rotundone, an Obscure Sesquiterpene, is a Potent Spicy Aroma Compound'. *Journal of Agricultural and Food Chemistry*, 56(10):, 3738–44.

Wysocki, C. J., & Beauchamp, G. K. (1984). 'Ability to Smell Androstenone is Genetically Determined'. *Proceedings of the National Academy of Sciences*, 81(15), 4899–902.

Chapter 12

Alexandre, H. (2013). 'Flor Yeasts of Saccharomyces Cerevisiae—Their Ecology, Genetics and Metabolism'. *International Journal of Food Microbiology*, 167(2), 269–75.

Boulton, R. B., Singleton, V. L., Bisson, L. F., Kunkee, R. E. (1999). 'The Role of Sulfur Dioxide in Wine'. In *Principles and Practices of Winemaking*, Springer New York (pp. 448–73).

Bueno, M., Carrascón, V., & Ferreira, V. (2016). 'Release and Formation of Oxidation-related Aldehydes During Wine Oxidation'. *Journal of Agricultural and Food Chemistry*, 64(3), 608–17.

Danilewicz, J. C. (2016). 'Fe (II):Fe (III) Ratio and Redox Status of White Wines'. *American Journal of Enology and Viticulture*, 67(2), 146–52.

Ebeler, S. E., & Spaulding, R. S. (1998). 'Characterization and Measurement of Aldehydes in Wine'. In Waterhouse, A. L. and Ebeler, S. E., *Chemistry of Wine Flavor*, ACS Publications (pp. 166–79).

Ferreira, V., Franco-Luesma, E., Vela, E., López, R., & Hernández-Orte, P. (2017). 'Elusive Chemistry of Hydrogen Sulphide and Mercaptans in Wine'. *Journal of Agricultural and Food Chemistry*, 66(10), 2237–46.

Killeen, D. J., Boulton, R., & Knoesen, A. (2018). 'Advanced Monitoring and Control of Redox Potential in Wine Fermentation'. *American Journal of Enology and Viticulture*, 69(4), 394–9.

Kilmartin, P. A. (2009). 'The Oxidation of Red and White Wines and its Impact on Wine Aroma'. *Chemistry in New Zealand*, 73(2), 79–83.

Linderholm, A. L., Findleton, C. L., Kumar, G., Hong, Y., & Bisson, L. F. (2008). 'Identification of Genes Affecting Hydrogen Sulphide Formation in Saccharomyces Cerevisiae'. *Applied and Environmental Microbiology*, 74(5), 1418–27.

Luckett, C. R., Pellegrino, R., Heatherly, M., et al. (2021). 'Discrimination of Complex Odour Mixtures: A Study Using Wine Aroma Models'. *Chemical Senses*, 46, bjaa079.

Morales, M. L., Ochoa, M., Valdivia, M., et al. (2020). 'Volatile Metabolites Produced by Different Flor Yeast Strains During Wine Biological Ageing'. *Food Research International*, 128, 108771.

Müller, N., & Rauhut, D. (2018). 'Recent Developments on the Origin and Nature of Reductive Sulphurous Off-odours in Wine'. *Fermentation*, 4(3), 62.

Naudé, Y., & Rohwer, E. R. (2013). 'Investigating the Coffee Flavour in South African Pinotage Wine Using Novel Offline Olfactometry and Comprehensive Gas Chromatography with

Time of Flight Mass Spectrometry'. *Journal of Chromatography A*, 1271(1), 176–80.

Ou, B., Huang, D., Hampsch-Woodill, M., & Flanagan, J. A. (2003). 'When East Meets West: The Relationship Between Yin-yang and Antioxidation-oxidation'. *The FASEB Journal*, 17(2), 127–9.

Schütz, M., & Kunkee, R. E. (1977). 'Formation of Hydrogen Sulphide from Elemental Sulphur During Fermentation by Wine Yeast'. *American Journal of Enology and Viticulture*, 28(3), 137–44.

Sharma, A., Kumar, R., Aier, I., et al. (2019). 'Sense of Smell: Structural, Functional, Mechanistic Advancements and Challenges in Human Olfactory Research'. *Current Neuropharmacology*, 17(9), 891–911.

Simpson, R. F. (2016). 'Aroma and Compositional Changes in Wine with Oxidation, Storage and Ageing'. *VITIS-Journal of Grapevine Research*, 17(3), 274.

Thanks

I am profoundly grateful to the numerous individuals who supported me throughout this journey. Here, I would like to express my appreciation for the people who have been vital contributors to my life, subsequently influencing the shaping of this book.

Foremost, I want to offer my heartfelt thanks to Dr Hildegarde Heymann for her inspiring lectures and insights into sensory science. The book she co-authored, *Sensory Evaluation of Foods: Principles and Practices*, is the most prescribed sensory textbook worldwide and it lays the foundational elements for the sensory aspects discussed in this book. I am also indebted to her for graciously reviewing the technical aspects of this book.

Special gratitude is also extended to Dr Andrew Waterhouse for his lucid guidance on wine chemistry and for entrusting me to co-author a review paper on wine pigments. His book *Understanding Wine Chemistry*, widely regarded as one of the finest works on the chemistry of wine, served as a profound inspiration for me as I embarked on writing this book. The encouragement of Drs Heymann and Waterhouse played a pivotal role in my ability to translate academic knowledge into papers and into this book, particularly given that English is not my native language.

Equally, I would like to express my sincere appreciation to other esteemed professors who significantly influenced my deep academic understanding of the world of wine: Dr Linda Bisson, Dr Roger Boulton, Dr Andy Walker, Dr David Block, Dr Dario Catu, and more.

To my dear classmates and friends in the grape growing and winemaking community, each of you with your unique perspectives on crafting the most satisfying wine, you continuously fuelled my desire to learn and explore. A special acknowledgement goes to Dr Gordon Walker, an expert and celebrity in the world of mushrooms, who provided invaluable advice during the most challenging times of writing this book. Philippe Venghiattis and Diane Wu, your kindness has provided me with precious moments of relaxation and enjoyment over the past seven years. To many other classmates from our time at UC Davis, you are the irreplaceable elements that coloured my adulthood.

I feel really lucky to work as a research and development scientist at the Harv 81 Group (Cork Supply & Tonnellerie Ô). Engaging in routine tasks involving chemical analysis and spearheading innovative sensory projects related to cork, oak and wine, I became well-versed in the chemical and sensorial terroir covered in this book. I am grateful for the opportunity to have a full-time job aligned with my deepest passions, propelling my professional growth. My sincere thanks go to my supervisor, Greg Hirson and my outstanding colleagues for their camaraderie and support, allowing me to focus on writing this book at home without undue stress.

Facilitated by the generosity of Peter Yeung, my connection to a reputable publisher specializing in wine publications, Académie du Vin Library, was made possible. Special thanks go to Rebecca Clare and Hermione Ireland of the Library for providing me with the opportunity to pen my inaugural book. Contributing to the esteemed legacy of Steven Spurrier, a highly revered and dearly missed figure in the industry, who founded the Library, is indeed a tremendous honour. Further acknowledgements are due to Ivy Xie and Julien Chen for their exceptional artwork showcased in this book.

Last but certainly not least, thanks to my parents for providing all possible resources for my education and career development. I am equally thankful for my extended family not bound by blood:

Fongyee Walker MW, Edward Ragg MW and Dr Huiqin Ma, who introduced me to the world of wine with unparalleled guidance and advice.

Cheers to each and every one of you!

Index